W0193268

Statistical Tools for Measuring Agreement

Lawrence Lin • A.S. Hedayat • Wenting Wu

Statistical Tools for Measuring Agreement

 Springer

Lawrence Lin
Baxter International Inc., WG3-2S
Rt. 120 and Wilson Rd.
Round Lake, IL 60073, USA
lawrence_lin@baxter.com

Wenting Wu
Mayo Clinic
200 First Street SW.
Rochester, MN 55905, USA
wu.wenting@mayo.edu

A.S. Hedayat
Department of Mathematics, Statistics
   and Computer Science
University of Illinois, Chicago
851 S. Morgan St.
Chicago, IL 60607-7045, USA
hedayat@uic.edu

ISBN 978-1-4614-0561-0   e-ISBN 978-1-4614-0562-7
DOI 10.1007/978-1-4614-0562-7
Springer New York Dordrecht Heidelberg London

Library of Congress Control Number: 2011935222

© Springer Science+Business Media, LLC 2012
All rights reserved. This work may not be translated or copied in whole or in part without the written
permission of the publisher (Springer Science+Business Media, LLC, 233 Spring Street, New York,
NY 10013, USA), except for brief excerpts in connection with reviews or scholarly analysis. Use in
connection with any form of information storage and retrieval, electronic adaptation, computer software,
or by similar or dissimilar methodology now known or hereafter developed is forbidden.
The use in this publication of trade names, trademarks, service marks, and similar terms, even if they are
not identified as such, is not to be taken as an expression of opinion as to whether or not they are subject
to proprietary rights.

Printed on acid-free paper

Springer is part of Springer Science+Business Media (www.springer.com)

To
    Sha-Li, Juintow, Buortau, and Shintau Lin
    Batool, Leyla, and Yashar Hedayat
    Xujian and MingEn Li

Preface

Agreement assessments are widely used in assessing the acceptability of a new or generic process, methodology and/or formulation in areas of lab performance, instrument/assay validation or method comparisons, statistical process control, goodness-of-fit, and individual bioequivalence. Successful applications in these situations require a sound understanding of both the underlying theory and practical problems in real life. This book seeks to blend theory and applications effectively and to present these two aspects with many practical examples.

The common theme in agreement assessment is to assess the agreement between observations of assay or rater (Y) and their target (reference) counterpart values (X). Target values may be considered random or fixed. Random target values are measured with random error. Common random target values are the gold standard of measurements, being both well established and widely acceptable. Sometimes we may also be interested in comparing two methods without a designated gold-standard method, or in comparing two technicians, times, reagents, or the like by the same method. Common fixed target values are the expected values or known values, which will be discussed in the most basic model presented in Chapters 2 and 3.

When there is a disagreement between methods, we need to know whether the source of the disagreement is due to a systematic shift (bias) or random error. Specific coefficients of accuracy and precision will be introduced to characterize these sources. This is particularly important in the medical-device environment, because a systematic shift usually can be easily fixed through calibration, while a random error usually is a more cumbersome variation-reduction exercise.

We will consider unscaled (absolute) and scaled (relative) agreement statistics for both continuous and categorical variables. Unscale agreement statistics are independent of between-sample variation, while the scale agreement statistics are relative to the between-sample variance. For continuous variables with proportional error, we often can simply apply a log transformation to the data and would evaluate percent changes rather than absolute differences. In practically all estimation cases, the statistical inference for parameter estimates will be discussed.

This book should appeal to a broad range of statisticians, researchers, practitioners, and students, in areas such as biomedical devices, psychology, and medical research in which agreement assessment is needed. Knowledge of regression, correlation, the asymptotic delta method, U-statistics, generalized estimation equations (GEE), and the mixed-effect model would be helpful in understanding the material presented and discussed in this book.

In Chapter 1, we will discuss definitions of precision, accuracy, and agreement, and discuss the pitfalls of some misleading approaches for continuous data.

In Chapter 2, we will start with the basic scenario of assessing agreement of two assays or raters, each with only one measurement for continuous data. In this basic scenario, we will consider the case of random or fixed target values for unscaled (absolute) and scaled (relative) indices with constant or proportional error structure.

In Chapter 3, we will introduce traditional approaches for categorical data with the basic scenario for unscaled and scaled indices. In terms of scaled agreement statistics, we will present the convergence of approaches for categorical and continuous data, and their association with a modified intraclass correlation coefficient. The information in this chapter and Chapter 2 sets the stage for discussing unified approaches in Chapters 5 and 6. In both Chapters 2 and 3, there is available a wealth of references to the basic model of agreement assessment. We will provide brief tours of related publications in these two chapters.

In Chapter 4, we will discuss sample size and power calculations for the basic models for continuous data. We will also introduce a simplified approach that is applicable to continuous and categorical data. We will present many practical examples in which we know only the most basic historical information such as residual variance or coefficient of variation.

In Chapter 5, we will consider a unified approach to evaluating agreement among multiple (k) raters, each with multiple replicates (m) for both continuous and categorical data. Under this general setting, intrarater precision, interrater agreement based on the average of m readings, and total-rater agreement based on individual readings will be discussed.

In Chapter 6, we will consider a flexible and general setting in which where the agreement of certain cases can be compared relative to the agreement of a chosen case. For example, to assess individual bioequivalence, we are interested in assessing the agreement of test and reference compounds relative to the agreement of the within-reference compound. As another example, in the medical-device environment, we often want to know whether the within-assay agreement of a newly developed assay is better than that of an existing assay. Both Chapters 5 and 6 are applicable to continuous and categorical data.

In Chapter 7, we will present a workshop using a continuous data set, a categorical data set, and an individual bioequivalence data set as examples. We will then address the use of SAS and R macros and the interpretation of the outputs from the most basic cases to more comprehensive cases.

This book is concise and concentrates on topics primarily based on the authors' research. However, proofs that were omitted from our published articles will be presented, and all other related tools will be well referenced. Many practical examples will be presented throughout the book in a wide variety of situations for continuous and categorical data.

A book such as this cannot have been written without substantial assistance from others. We are indebted to the many contributors who have developed the theory and practice discussed in this book. We also would like to acknowledge our appreciation of the students at the University of Illinois at Chicago (UIC) who helped us in many ways. Specifically, six PhD dissertations on agreement subjects have been produced by Robieson (1999), Zhong (2001), Yang (2002), Wu (2005), Lou (2006) and Tang (2010). Their contributions have been the major sources for this book. Most of the typing using MikTeX was performed by the UIC PhD student Mr. Yue Yu, who also double-checked the accuracy of all the formulas.

We would like to mention that we have found the research into theory and application performed by Professors Tanya King, of the Pennsylvania State Hershey College of Medicine; Vernon Chinchilli, of the Pennsylvania State University College of Medicine; and Huiman Barnhart, of the Duke Clinical Research Institute, are truly inspirational. Their work has influenced our direction for developing the materials of our book. We are also indebted to Professor Phillip Schluter, of the School of Public Health and Psychosocial Studies at AUT University, New Zealand, for his permission to use the data presented in Examples 5.9.3 and 6.7.2 prior to their publication.

Finally, all SAS and R macros and most data in the examples are provided at the web sites shown below:

1. http://www.uic.edu/~hedayat/
2. http://mayoresearch.mayo.edu/biostat/sasmacros.cfm

The U.S. National Science Foundation supported this project under Grants DMS-06-03761 and DMS- 09-04125.

Round Lake, IL, USA   Lawrence Lin
Chicago, IL, USA   Samad Hedayat
Rochester, MN, USA   Wenting Wu

Contents

Symbols Used and Abbreviations

In this book, we use a Greek letter (symbol) to represent a parameter to be estimated, and we use its respective English letter or the symbol with a hat to represent its sample counterpart or estimate. The exception is that we use $\overline{X}$ to represent the sample mean, due to the long history of that convention. When a transformation is performed, we use an uppercase letter to represent a transformed estimate. However, there are some complicated computational formulas in which we use uppercase letters to simplify the computation. In the sequel, we use Greek letters to represent parameters when the target value X is considered random. When the target value is considered fixed, we add $|X$ as a subscript to the corresponding parameter. For example, $\varepsilon^2_{|X}$ represents the mean squared deviation (MSD) when the target value X is assumed fixed. We use a boldface symbol or letter to represent a vector or matrix. Symbols and their corresponding definitions are listed below:

ε^2	Mean squared deviation
δ_{π_0}	Total deviation index
π_{δ_0}	Coverage probability
ρ_c	Concordance correlation coefficient
υ	Location shift
ϖ	Scale shift
χ_a	Accuracy coefficient
ρ	Precision coefficient
κ	Kappa
κ_w	Weighted kappa
ζ	Total-rater MSD to Intra-rater MSD ratio
ψ	Intra-rater MSD to Intra-rater MSD ratio
Δ	Relative bias squared

Abbreviations used in this book are (in alphabetical order):

CCC:	Concordance correlation coefficient
CDF:	Cumulative density function
CIA:	Coefficient of individual agreement
CL:	Confidence limit

CLIA: Clinical laboratory improvement amendments
CP: Coverage probability
GEE: Generalized estimation equations
GM: Geometric mean
ICC: Intraclass correlation coefficient
IIR: Intra rater MSD to Intra rater MSD ratio
ML: Maximum likelihood
MLE: Maximum likelihood estimate
MSD: Mean squared deviation
PT: Proficient testing
PTC: Proficient testing criterion
RBS: Relative bias squared
RML: Restricted maximum likelihood
RMLE: Restricted maximum likelihood estimate
SD: Standard deviation
TDI: Total deviation index based on absolute difference
TDI%:   Total deviation index based on percent change
TIR: Total-rater MSD to Intra rater MSD ratio

Chapter 1
Introduction

Consider the problems of assessing the acceptability of a new or generic process, methodology, and/or formulation in areas of lab performance, instrument/assay validation or method comparisons, statistical process control, goodness-of-fit, and individual bioequivalence. The common theme is to assess the agreement between observations (Y) and their corresponding target values (X). Target values may be considered random or fixed. Commonly used random target values are the gold standard measurements, which are proven and widely acceptable. Commonly used fixed target values are the expected or known values. We might be interested in comparing two methods without a designated gold standard method. Sometimes, we may also be interested in comparing a newly developed assay that is alleged to be more precise and accurate than a designated gold standard assay. Within a method, we might be interested in comparing technicians/times/reagents.

For simplicity and the ease of reference, we will use the term *assays* and *raters* to represent assays, raters, instruments, methods, etc. Also, we will use the term *samples* to designate samples, patients, animals, or subjects. In the tradition of the subject matter, we use throughout this book the terms *index* and *coefficient* interchangeably.

Figure 1.1 presents a typical situation for assessing agreement. When we plot the observed values on the y-axis versus the corresponding target values over a desirable range on the x-axis, we would like to see agreement in the paired data so that the observations fall closely along the identity line, which is the straight line with zero intercept and unit slope. When there is evidence of disagreement, it is important to address the issue and search for the sources of that disagreement.

1.1 Precision, Accuracy, and Agreement

Generally, the common basic sources of disagreement come from within-sample variation (imprecision) and/or a shift in the marginal distributions (inaccuracy). Fixing imprecision is a within-sample variance reduction exercise in the medical-device

L. Lin et al., *Statistical Tools for Measuring Agreement*,
DOI 10.1007/978-1-4614-0562-7_1, © Springer Science+Business Media, LLC 2012

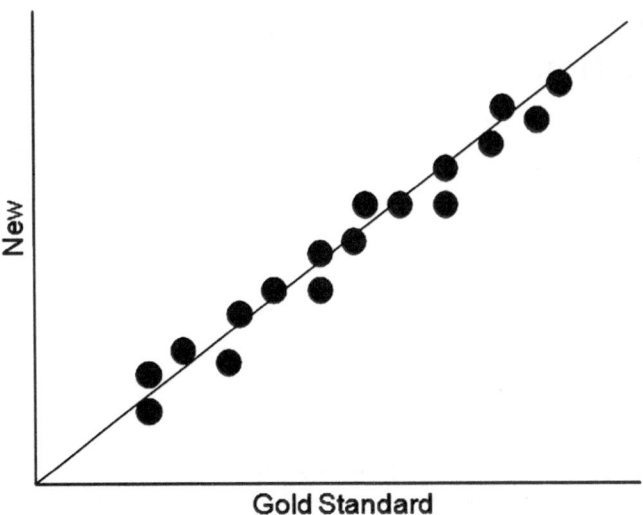

Fig. 1.1 Assessing agreement of observed values (new) and target values (gold standard method)

or engineering environment, which is typically more cumbersome than fixing inaccuracy. The cause of inaccuracy (systematic bias) is most likely a calibration problem in measuring devices. In the published literature and regulatory documents, we find the following definitions for precision, accuracy, and agreement.

The Food and Drug Administration (FDA) guidelines on bioanalytical method validation (http://www.fda.gov/downloads/Drugs/GuidanceComplianceRegulatory Information/Guidances/UCM070107.pdf) defines accuracy as deviation of mean from true value (trueness), while the International Organization for Standardization (ISO) 5725 (1994) defines accuracy as both "trueness" and "precision." When there is no hard true value, we can loosely define improved accuracy as less bias.

The FDA defines precision as "the closeness of agreement (degree of scatter) between a series of measurements obtained from multiple sampling of the same homogeneous sample under the prescribed conditions" while the ISO defines precision as the closeness of agreement between independent test results obtained under **stipulated** conditions measures from within-sample variation. The FDA further defines "precision under tight conditions (intra-batch, inter-batch, true replicates)" as repeatability, and "precision across labs" as reproducibility. Other precision definitions could include intraclass correlation coefficient (ICC) for measuring reliability in the social sciences.

Agreement is often defined as having both precision and accuracy, which is a function of the absolute difference between pair readings. There is confusion in the above definitions. For example, the ISO's definition of accuracy is confused with the definition of agreement. We adopt the accuracy and precision that correspond to the FDA definitions. There is a host of terminology related to precision such as repeatability, reproducibility, robustness, validity, and reliability. We shall try to keep the terminology simple and straightforward, such as agreement/precision/accuracy across assays/days/technicians/labs.

1.2 Traditional Approaches for Continuous Data

Traditionally, the agreement between observed and target values has been assessed by a paired t-test, the results of which could be misleading. A slightly better approach in the assessment of agreement is to test the linear least squares estimates against the identity line (a straight line with zero intercept and unit slope). We will denote this method by LS01. Furthermore, the target values have been assumed fixed in the LS01 (regression) analysis, even when the target values are obviously random. These two rudimentary approaches capture only the accuracy information relative to the precision. The potential for obtaining misleading results using these two approaches can be illustrated graphically by Fig. 1.2.

Another popular traditional method for assessing agreement of observed and target values is the Pearson correlation coefficient. However, this coefficient is only a measure of precision. The potential for obtaining misleading results using this approach can be illustrated by Fig. 1.3. In addition, this coefficient has frequently been misused to assess linearity, which should have been assessed through goodness-of-fit statistics. Other traditionally used approaches for assessing agreement of observed and target values have included the coefficient of variation and the mean square error of the regression analysis, which are measures of precision only.

Perhaps the most valid traditional approach for assessing agreement of observed and target values has been the intraclass correlation coefficient (ICC). The ICC in its original form (Fisher 1925) is the ratio of between-sample variance and total (within + between) variance under the model of equal marginal distributions. This original ICC was intended to measure precision only. This coefficient is invariant with respect to the interchanges of Y and X values within any pairs. Several forms of ICC have evolved. In particular, Bartko (1966), Shrout and Fleiss (1979), Fleiss (1986), and Brennan (2001) have put forth various reliability assessments. We will discuss ICC in greater detail in Section 2.3.1 and Chapter 3. In Chapter 5, we will

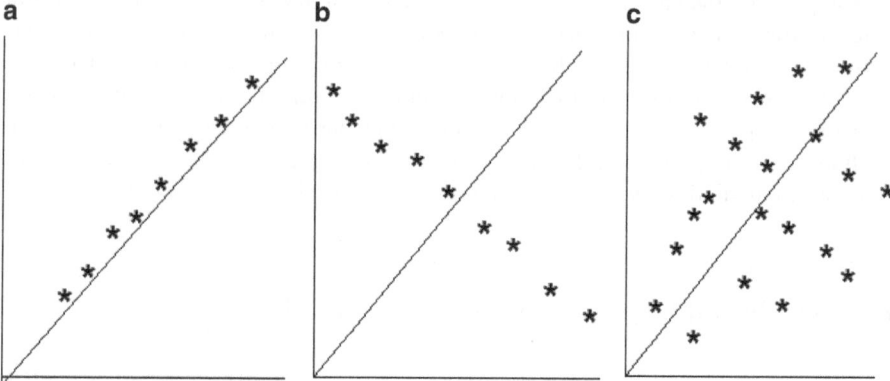

Fig. 1.2 Situations in which a paired t-test or least squares test against the identity line (LS01) can be misleading: (**a**) Rejected by paired t-test/LS01. (**b**) Accepted by paired t-test but rejected by LS01. (**c**) Accepted by paired t-test/LS01

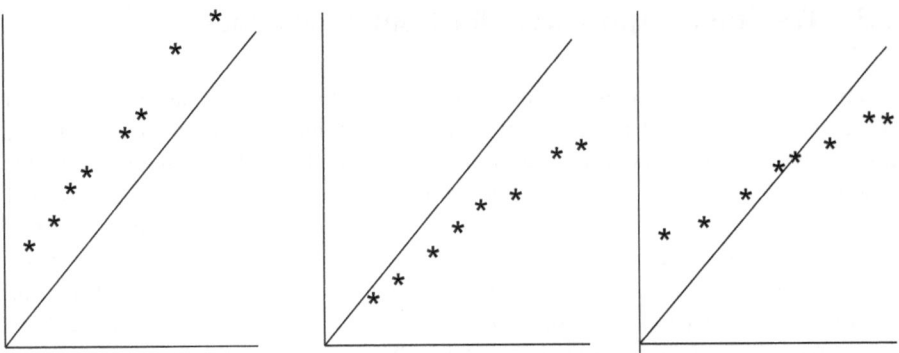

Fig. 1.3 Situations in which the Pearson correlation coefficient can be misleading

introduce some special forms of ICC under the general case that correspond to agreement, precision, and accuracy for both continuous and categorical data, based on a paper by Lin, Hedayat, and Wu (2007).

It should be pointed out that the traditional hypothesis-testing approach is not appropriate for assessing agreement except for the cases that will be presented in Chapter 6. In traditional hypothesis testing, the rejection region (alternative hypothesis) is the region for declaring a difference based on strong evidence presented in the data. Failing to reject the null hypothesis does not imply accepting agreement, but implies a lack of evidence for declaring a difference. The proper setting for assessing agreement is to reverse the null and alternative hypotheses, so that the conventional rejection region actually is the region for declaring agreement. Therefore, we would reject the null hypothesis of a difference and accept the alternative hypothesis of agreement based on strong evidence presented in the data (Dunnett and Gent 1977; Bross 1985; Rodary, Com-Nougue, and Tournade 1989; Lin 1992). With the given criterion and the same precision of the data, the larger the sample size, the easier it should be to accept the agreement. Here, a meaningful criterion for an acceptable difference should be prespecified, and the hypothesis testing should be one-sided. Indeed, the proper hypothesis-testing approach is equivalent to computing the one-sided confidence limit. If this limit were better than the prespecified criterion, we would accept the agreement. We will use the confidence limit approach in this book for simplicity. However, for the sample size and power calculation, we will use the proper hypothesis-testing approach.

1.3 Traditional Approaches for Categorical Data

An example of categorical response is for two or more assays to assign a sample's condition according to a binary scale of no/yes or normal/not normal, or to an ordinal scale of fair, mild, serious, critical, or life-threatening. A nonordinal (nominal) scale in assessing agreement is less often encountered in practical situations.

Compared to approaches for continuous data, there have been fewer misleading approaches for categorical data. The most popular approach for assessing agreement began with kappa (Cohen 1960) and weighted kappa (Cohen 1968; Fleiss, Cohen, and Everitt 1969). The kappa coefficients assess nonchance (chance-corrected) agreement relative to the total nonchance agreement. There is a long history of valid tools available for assessing marginal equivalence, association, and agreement. These will be referenced in Chapter 3.

Chapter 2
Continuous Data

We will now introduce new approaches that have evolved for measuring agreement since 1988. Some of these new approaches were summarized, studied, and compared by Lin, Hedayat, Sinha, and Yang (2002). Here, we include the necessary proofs that were left out of that article. In addition, we include assorted examples to demonstrate agreement techniques. We begin with the most basic model, in which paired observations (Y and X) are collected.

2.1 Basic Model

When target values are random, the joint distribution of Y and X is assumed to have a bivariate distribution with finite second moments with means μ_y, μ_x, variances σ_y^2, σ_x^2 and covariance σ_{yx}. When target values are fixed, $Y_i|X_i$, $i = 1,\ldots,n$, are assumed to be observations in a random sample from the basic regression model $Y = \beta_0 + \beta_1 X + e_Y$. Here, e_Y is the residual error with mean 0 and variance σ_e^2.

2.2 Absolute Indices

2.2.1 Mean Squared Deviation

Mean squared deviation (MSD) evaluates an aggregated deviation from the identity line, $\text{MSD} = E(Y - X)^2$. It can be expressed as

$$\varepsilon^2 = (\mu_y - \mu_x)^2 + \sigma_y^2 + \sigma_x^2 - 2\sigma_{yx}, \tag{2.1}$$

when target values are random, or

$$\varepsilon_{|X}^2 = (\mu_{y|\overline{X}} - \overline{X})^2 + s_x^2(1 - \beta_1)^2 + \sigma_e^2, \tag{2.2}$$

L. Lin et al., *Statistical Tools for Measuring Agreement*,
DOI 10.1007/978-1-4614-0562-7_2, © Springer Science+Business Media, LLC 2012

when target values are fixed, where $\overline{X}$ and s_x^2 are the sample mean and variance of X.

Estimated by sample counterparts (e^2 or $e_{|X}^2$) with a log transformation, $W = \ln(e^2)$ or $W_{|X} = \ln(e_{|X}^2)$ has an asymptotic normal distribution with mean $\ln(\varepsilon^2)$ or $\ln(\varepsilon_{|X}^2)$ and variance

$$\sigma_W^2 = \frac{2}{n-2}\left[1 - \frac{(\mu_y - \mu_x)^4}{\varepsilon^4}\right], \tag{2.3}$$

when target values are random, or

$$\sigma_{W|X}^2 = \frac{2}{n-2}\left[1 - \frac{\left(\varepsilon_{|X}^2 - \sigma_e^2\right)^2}{\varepsilon_{|X}^4}\right], \tag{2.4}$$

when target values are fixed.

The proof of (2.3) can be found in Section 2.9.2. The proof of (2.4) can be found in Section 2.10.2. For statistical inference, refer to Section 2.4 for the information regarding sample counterparts, and to Sections 2.9 and 2.10 for the proofs of asymptotic normality in this chapter.

The MSD is not an easy index to interpret. The following methods will put some meaningful interpretation on this basic MSD index.

2.2.2 Total Deviation Index

To examine the agreement from a different perspective, a measure that captures a large proportion (π_0) of data within a boundary (δ_0) from target values was considered by Lin, Hedayat, Sinha, and Yang (2002). For example, we may want to capture at least 90% of individual observations that are within 10% of their target values. We would compute total deviation index (TDI) for the given coverage probability (CP) criterion of 0.9 to see whether this TDI is less than 10%, or compute coverage probability (CP) for the given TDI criterion of 10% to see whether this CP is more than 0.9.

Assume that $D = Y - X$ has a normal distribution with mean $\mu_d = \mu_y - \mu_x$ and variance $\sigma_d^2 = \sigma_y^2 + \sigma_x^2 - 2\sigma_{xy}$. We may find π for a given δ_0 criterion, CP_{δ_0}, which is

$$\pi_{\delta_0} = P(D^2 < \delta_0^2) = \chi^2\left(\delta_0^2, 1, \frac{\mu_d^2}{\sigma_d^2}\right), \tag{2.5}$$

where $\chi^2(\cdot)$ is the cumulative noncentral chi-square distribution up to δ_0^2, with one degree of freedom and noncentrality parameter $\frac{\mu_d^2}{\sigma_d^2}$. This measure will be presented shortly.

We may also find δ for a given π_0 criterion, TDI_{π_0}, which is

$$\delta_{\pi_0} = \sqrt{(\chi^2)^{-1}\left(\pi_0, 1, \frac{\mu_d^2}{\sigma_d^2}\right)}, \tag{2.6}$$

where $(\chi^2)^{-1}$ is the inverse function of $\chi^2(\cdot)$. Since the estimate of this index has intractable asymptotic properties, Lin (2000) and Lin, Hedayat, Sinha, and Yang (2002) have suggested the following TDI_{π_0} approximation:

$$\delta_{\pi_0\sim}^2 \doteq (\chi^2)^{-1}(\pi_0, 1)\varepsilon^2, \tag{2.7}$$

or

$$\delta_{\pi_0\sim} \doteq \Phi^{-1}\left(1 - \frac{1 - \pi_0}{2}\right)|\varepsilon|, \tag{2.8}$$

when X is random, or

$$\delta_{\pi_0\sim|X} \doteq \Phi^{-1}\left(1 - \frac{1 - \pi_0}{2}\right)|\varepsilon_{|X}|, \tag{2.9}$$

when X is fixed.

The approximation is satisfactory (Lin 2000) when:

1. $\pi_0 = 0.75$ and $\Delta \leq 1/2$,
2. $\pi_0 = 0.8$ and $\Delta \leq 8$,
3. $\pi_0 = 0.85$ and $\Delta \leq 2$,
4. $\pi_0 = 0.9$ and $\Delta \leq 1$,
5. $\pi_0 = 0.95$ and $\Delta \leq 1/2$.

The quantity $\Delta = \frac{\mu_d^2}{\sigma_d^2}$ is called the *relative bias squared* (RBS). The interpretation of this approximated TDI is that approximately $100\pi_0\%$ of observations are within $\delta_{\pi_0\sim}$ of the target values. $\text{TDI}_{\pi_0}^2$ is proportional to MSD, and therefore we may perform an inference based on the asymptotic normality of $W = \ln(e^2)$, where e^2 is the sample counterpart of MSD when X is random, or $W_{|X} = \ln(e_{|X}^2)$ when X is fixed. This simplified method will become very useful when we deal with the more general case to be introduced in Chapter 5.

The idea of using such an approximation of TDI was motivated by Holder and Hsuan (1993). They proposed a moment-based criterion for assessing individual bioequivalence. They showed that in a slightly different fashion, $\delta_{\pi_0}^2$, or the squared function of (2.6), has an upper bound $\delta_{\pi_0+}^2 = c_{\pi_0}\varepsilon^2$, where c_{π_0} is a constant not depending on μ_d and σ_d. Therefore, δ_{π_0+} conservatively captures at least $100\pi_0\%$ of observations within the boundary from target values of a reference compound. Holder and Hsuan (1993) used a numerical algorithm for the determination of c_{π_0} under some parametric and nonparametric distribution of $D = Y - X$.

However, the asymptotic distribution property of this estimate has not been established. Lin (2000) made a comparison between this statistic and the TDI given in (2.7). When $\pi_0 = 0.9$, $\delta^2_{\pi_0\sim}$ and $\delta^2_{\pi_0+}$ are identical under the normality assumption. Using $\text{TDI}_{0.9}$ is almost exact when $\frac{\mu_d}{\sigma_d} < 1$, and would become conservative otherwise. Using $\text{TDI}_{0.8}$ is most robust, since it can tolerate an RBS value as high as 8.0.

A TDI is similar in concept to a tolerance limit. The difference is that a tolerance limit captures individual deviations from their own mean, while a TDI captures individual deviations from their target values, for a high proportion (say, 90%), and with a high degree of confidence (say, 95%) when the upper confidence limit of TDI is used.

2.2.3 Coverage Probability

We now consider finding π for a given δ_0 criterion. This is

$$\pi_{\delta_0} = P(D^2 < \delta_0^2) = \chi^2\left(\delta_0^2, 1, \frac{\mu_d^2}{\sigma_d^2}\right), \tag{2.10}$$

when target values are random, or

$$\pi_{\delta_0|X} = \frac{1}{n}\sum_{i=1}^{n}\pi_{\delta_0(i)}, \tag{2.11}$$

where

$$\pi_{\delta_0(i)} = \chi^2\left[\frac{\delta_0^2}{\sigma_e^2}, 1, \left(\frac{\beta_0 + (\beta_1 - 1)X_i}{\sigma_e}\right)^2\right], \tag{2.12}$$

when target values are fixed. The estimate of coverage probability using sample counterparts (p_{δ_0} or $p_{\delta_0|X}$) by the logit transformation, $T = \ln\frac{p_{\delta_0}}{1-p_{\delta_0}}$ or $T_{|X} = \ln\frac{p_{\delta_0|X}}{1-p_{\delta_0|X}}$, has an asymptotic normal distribution with mean $E(p_{\delta_0}) = \ln\frac{\pi_{\delta_0}}{1-\pi_{\delta_0}}$ or $E(p_{\delta_0|X}) = \ln\frac{\pi_{\delta_0|X}}{1-\pi_{\delta_0|X}}$, and variance

$$\sigma_T^2 = \frac{0.5\left[\delta_{+\mu\sigma}\phi(-\delta_{+\mu\sigma}) + \delta_{-\mu\sigma}\phi(\delta_{-\mu\sigma})\right]^2 + \left[\phi(-\delta_{+\mu\sigma}) - \phi(\delta_{-\mu\sigma})\right]^2}{(n-3)(1-\pi_{\delta_0})^2\pi_{\delta_0}^2}, \tag{2.13}$$

when X is random, or

$$\sigma_{T|\overline{X}}^2 = \frac{\left[\frac{C_0^2}{n^2} + \frac{(C_0\overline{X}-C_1)^2}{n^2 s_x^2} + \frac{C_2^2}{2n^2}\right]}{(n-3)(1-\pi_{\delta_0})^2\pi_{\delta_0}^2}, \tag{2.14}$$

when X is fixed, where

$$\delta_{+\mu\sigma} = \frac{\delta_0 + \mu_d}{\sigma_d},\tag{2.15}$$

$$\delta_{-\mu\sigma} = \frac{\delta_0 - \mu_d}{\sigma_d},\tag{2.16}$$

$$\delta_{+\mu\beta\sigma_i} = \frac{\delta_0 + \beta_0 + (\beta_1 - 1)X_i}{\sigma_e},\tag{2.17}$$

$$\delta_{-\mu\beta\sigma_i} = \frac{\delta_0 - \beta_0 - (\beta_1 - 1)X_i}{\sigma_e},\tag{2.18}$$

$$C_0 = \sum_{i=1}^{n} \left[\phi(-\delta_{+\mu\beta\sigma_i}) - \phi(\delta_{-\mu\beta\sigma_i})\right],\tag{2.19}$$

$$C_1 = \sum_{i=1}^{n} \left[\phi(-\delta_{+\mu\beta\sigma_i}) - \phi(\delta_{-\mu\beta\sigma_i})\right] X_i,\tag{2.20}$$

$$C_2 = \sum_{i=1}^{n} \left[\delta_{+\mu\beta\sigma_i}\phi(-\delta_{+\mu\beta\sigma_i}) + \delta_{-\mu\beta\sigma_i}\phi(\delta_{-\mu\beta\sigma_i})\right],\tag{2.21}$$

and $\phi(\cdot)$ is the standard normal distribution function.

The proof of (2.13) can be found in Section 2.9.3. The proof of (2.14) can be found in Section 2.10.3.

We can also use the simplified method of

$$\pi_{\delta_0\sim} \doteq \chi^2\left(\frac{\delta_0^2}{\varepsilon^2}, 1\right),\tag{2.22}$$

which will become very useful when we deal with the more general case to be introduced in Chapter 5.

2.3   Relative Indices

2.3.1   Intraclass Correlation Coefficient

Pearson (1899, 1901) developed the intraclass correlation coefficient (ICC) as a way to estimate various aspects of fraternal resemblance. Given pairs of brothers, one might be interested not in the correlation in height between the older and the younger brother, or the taller and the shorter brother, but simply between brothers in general. In this case, the heights of the two brothers are logically interchangeable. Pearson suggested that this correlation could be estimated by entering the height measurements for each pair of brothers, (x, y), twice into the computation of the usual product–moment correlation coefficient ρ, once in the order (x, y) and once in the order (y, x). If there are more than two brothers in each set, each possible pair of measurements is entered twice into the computation. Thus, the number of entries in the correlation for this set is $n(n - 1)$, where n is the number of brothers in the data set.

Harris (1913) developed a simple formula for intraclass correlation as a function of (a) the variance of the means of each set of measurements around the overall mean and (b) the variance of the total set of measurements.

Fisher (1925) observed that variance measurements could be partitioned into two components. The first component is the between-sample variance after removing the residual variance, which Fisher called A. The second component is the residual variance or within-sample variance, which Fisher called B. Thus the population intraclass correlation can be expressed as

$$\rho_I = \frac{A}{A + B}. \tag{2.23}$$

Fisher (1925) noted that the ICC could be estimated using mean squares from an analysis of variance (ANOVA). We will revisit ICC in Chapter 3, where we will show its association with kappa, weighted kappa, and the concordance correlation coefficient (CCC) presented below. We will also revisit, in Chapter 5, the general form of the ICC for agreement, precision, and accuracy coefficients.

2.3.2 Concordance Correlation Coefficient

The MSD can be standardized such that: 1 indicates that each pair of readings is in perfect agreement in the population (for example, 1, 1; 2, 2; 3, 3; 4, 4; 5, 5), while 0 indicates no correlation, and -1 means that each pair of readings is in perfect reversed agreement in the population (for example, 5, 1; 4, 2; 3, 3; 2, 4; 1, 5). Lin (1989) introduced one such standardization of MSD, called CCC, which is defined as

$$\rho_c = 1 - \frac{\varepsilon^2}{\varepsilon^2_{|\rho=0}} \tag{2.24}$$

$$= 1 - \frac{\varepsilon^2}{\sigma_y^2 + \sigma_x^2 + (\mu_y - \mu_x)^2}$$

$$= \frac{2\sigma_{yx}}{\sigma_y^2 + \sigma_x^2 + (\mu_y - \mu_x)^2},$$

$$= \frac{2\rho\sigma_x\sigma_y}{\sigma_y^2 + \sigma_x^2 + (\mu_y - \mu_x)^2} \tag{2.25}$$

when X is random, and

$$\rho_{c|X} = \frac{2\beta_1 s_x^2}{\sigma_y^2 + s_x^2 + (\mu_y - \bar{X})^2},$$ (2.26)

when X is fixed. The CCC is closely related to the intraclass correlation and has a meaningful geometrical interpretation. It is inversely related to the mean square of the ratio of the within-sample total deviation (ε^2) and the total deviation $(\varepsilon^2_{|\rho=0})$. For example, if the within-sample total deviation is 10%, 32%, or 45% of the total deviation, then the CCC is $0.99 = (1 - 0.1^2)$, $0.90 = (1 - 0.32^2)$, or $0.80 = (1 - 0.45^2)$, respectively. In Chapter 3, we will show that for ordinal categorical data, CCC degenerates into the weighted kappa suggested by Cohen (1968).

Section 1.1 defined accuracy and precision in the one-dimensional situation. According to the two-dimensional model of Section 2.1, the between-sample variation is typically inherited or is a result of the design of the sampling process, and is usually unrelated to within-sample precision of an assay. Therefore, we consider the difference in between-sample variance as a systematic bias, and it is included in the inaccuracy. A sample mean and sample variance define a marginal distribution in most of the commonly used distributions.

2.3.2.1   Accuracy Coefficient

The accuracy coefficient measures the closeness of the marginal distributions of Y and X, where 1 signifies equal means and variances, and 0 indicates that the absolute difference in means and/or variance approach infinity. The accuracy coefficient can be broken down into measures of location and/or scale shifts, where the location shift is $\upsilon = \frac{\mu_y - \mu_x}{\sqrt{\sigma_y \sigma_x}}$, and the scale shift is $\varpi = \frac{\sigma_y}{\sigma_x}$ or $\frac{\sigma_x}{\sigma_y}$. Here, the accuracy coefficient is defined as

$$\chi_a = \frac{2}{\varpi + 1/\varpi + \upsilon^2},$$ (2.27)

when X is random. We can replace σ_x^2 by s_x^2, and μ_x by $\bar{X}$ in (2.27) when X is fixed.

2.3.2.2   Precision Coefficient

The precision coefficient is the Pearson correlation coefficient (ρ) between Y and X, where

$$\rho = \frac{\sigma_{yx}}{\sigma_y \sigma_x}.$$ (2.28)

Here ρ^2 has the same scale as the accuracy coefficient, from 0 (no agreement) to 1 (perfect agreement). It is evident from (2.25), (2.26), and (2.27) that the CCC is the product of the precision and accuracy coefficients when target values are random or fixed.

2.3.2.3 Statistical Inference on CCC and the Accuracy and Precision Coefficients

The estimate of CCC (r_c or $r_{c|X}$) using sample counterparts by the Z transformation, $Z = \frac{1}{2}\ln\frac{1+r_c}{1-r_c}$ or $Z_{|X} = \frac{1}{2}\ln\frac{1+r_{c|X}}{1-r_{c|X}}$, has an asymptotic normal distribution with mean $\frac{1}{2}\ln\frac{1+\rho_c}{1-\rho_c}$ or $\frac{1}{2}\ln\frac{1+\rho_{c|X}}{1-\rho_{c|X}}$, and variance

$$\sigma_Z^2 = \frac{1}{(n-2)}\left[\frac{(1-\rho^2)\rho_c^2}{(1-\rho_c^2)\rho^2} + \frac{2\rho_c^3(1-\rho_c)\upsilon^2}{\rho(1-\rho_c^2)^2} - \frac{\rho_c^4\upsilon^4}{2\rho^2(1-\rho_c^2)^2}\right] \tag{2.29}$$

or

$$\sigma_{Z|X}^2 = \frac{(1-\rho^2)\rho_c^2}{[(n-2)(1-\rho_c^2)^2\rho^2]}\left[\varpi\upsilon^2\rho_c^2 + (1-\rho_c\rho\varpi)^2 + \frac{\varpi^2\rho_c^2(1-\rho^2)}{2}\right], \tag{2.30}$$

when X is random or fixed, respectively.

The estimate of the accuracy coefficient (c_a or $c_{a|X}$) using sample counterparts by the logit transformation, $L = \ln\frac{c_a}{1-c_a}$ or $L_{|X} = \ln\frac{c_{a|X}}{1-c_{a|X}}$, has an asymptotic normal distribution with mean $\ln\frac{\chi_a}{1-\chi_a}$ or $\ln\frac{\chi_{a|X}}{1-\chi_{a|X}}$, and variance

$$\sigma_L^2 = \frac{\chi_a^2\upsilon^2(\varpi+1/\varpi-2\rho) + \frac{1}{2}\chi_a^2(\varpi^2+1/\varpi^2+2\rho^2) + (1+\rho^2)(\chi_a\upsilon^2-1)}{(n-2)(1-\chi_a)^2} \tag{2.31}$$

or

$$\sigma_{L|X}^2 = \frac{\upsilon^2\varpi\chi_a^2(1-\rho^2) + \frac{1}{2}(1-\varpi\chi_a)^2(1-\rho^4)}{(n-2)(1-\chi_a)^2}. \tag{2.32}$$

The estimate of the precision coefficient (r or $r_{|X}$) by the Z-transformation has an asymptotic normal distribution with mean $\frac{1}{2}\ln\frac{(1+\rho)}{(1-\rho)}$ or $\frac{1}{2}\ln\frac{(1+\rho_{|X})}{(1-\rho_{|X})}$, and variance $\frac{1}{n-3}$ when target values are random, or $\frac{1-\rho^2/2}{n-3}$ when target values are fixed. The proof of (2.29) and (2.30) can be found in Sections 2.9.1 and 2.10.1. The proof of (2.31) and (2.32) can be found in Sections 2.9.4 and 2.10.4.

2.4 Sample Counterparts

For the purpose of statistical inference, the parameters discussed above can be replaced with their sample estimates that are consistent estimators, such as the moments estimates.

The sample counterparts for μ_y, μ_x, σ_y^2, σ_x^2, σ_{yx}, β_1, and ρ are

$$\overline{Y} = \frac{1}{n}\sum_{i=1}^{n} Y_i, \tag{2.33}$$

$$\overline{X} = \frac{1}{n}\sum_{i=1}^{n} X_i, \tag{2.34}$$

$$s_y^2 = \frac{1}{n}\sum_{i=1}^{n}(Y_i - \overline{Y})^2, \tag{2.35}$$

$$s_x^2 = \frac{1}{n}\sum_{i=1}^{n}(X_i - \overline{X})^2, \tag{2.36}$$

$$s_{yx}^2 = \frac{1}{n}\sum_{i=1}^{n}(Y_i - \overline{Y})(X_i - \overline{X}), \tag{2.37}$$

$$b_1 = \frac{s_{yx}}{s_x^2}, \tag{2.38}$$

and

$$r = \frac{s_{yx}}{s_y s_x}. \tag{2.39}$$

We could certainly use $\frac{1}{n-1}$ rather than $\frac{1}{n}$ in the above variance and covariance estimates. However, we use $\frac{1}{n}$ to bound the CCC estimate by 1.0.

In estimating the MSD, we use the sum of squared difference divided by $n-1$. For (2.17) and (2.18) when X is fixed, we use

$$s_e^2 = \frac{n}{(n-3)}(1 - r^2)s_y^2. \tag{2.40}$$

For less bias in estimating the RBS, and in estimating σ_d^2 in (2.15) and (2.16), we use

$$s_d^2 = \frac{n}{(n-3)}\left(s_x^2 + s_y^2 - 2s_{xy}\right). \tag{2.41}$$

The use of $n-2$ or $n-3$ instead of n in the denominators of the above variance equations is for the small-sample-size bias correction based on the simulation studies in Lin, Hedayat, Sinha, and Yang (2002). The use of sample-size bias correction is not important when the sample size is large.

For the purpose of performing statistical inference for each index, we should compute the confidence limit (lower limit for CCC, precision coefficient, accuracy coefficient, CP, and upper limit for TDI) based on its respective transformation, then perform antitransformation to the limit. We will declare that the assay agreement is acceptable when the limit is better than the prespecified criterion. The use of transformed estimates can speed up the approach to normality. Moreover, a transformation could bound the confidence interval to its respective parameter range, say, -1 to 1 for CCC and precision coefficient, 0 to 1 for the accuracy coefficient and CP, and 0 to infinity for MSD.

Throughout this book, once the asymptotic normality of an estimated index has been defined, statistical inference can be established through confidence limit(s). Let $\hat{\lambda}$ be the estimate of an agreement index and $\sigma_{\hat{\lambda}}^2$ its variance. Then the one-sided upper or lower confidence limit becomes

$$\hat{\lambda} + \Phi^{-1}(1-\alpha)\hat{\sigma}_{\hat{\lambda}} \quad \text{or} \quad \hat{\lambda} - \Phi^{-1}(1-\alpha)\hat{\sigma}_{\hat{\lambda}},$$

where $\hat{\sigma}_{\hat{\lambda}}$ is the estimate of the square root of variance, $\sigma_{\hat{\lambda}}$, using sample counterparts. When the sample size is small, say less than 30, we can also use the cutoff value of the cumulative central t-distribution instead of the standard cumulative normal distribution to form the statistical inference.

2.5 Proportional Error Case

When Y and X are positively valued variables and the standard deviations of Y are proportional to either Y or X, it is assumed that $\ln(Y)$ and $\ln(X)$ have a bivariate normal distribution. Let $100\theta\%$ be the percent change in Y and X. Then

$$\pi_\theta = P\left[\frac{1}{(1+\theta)} < \frac{Y}{X} < (1+\theta)\right] = P\left[|\ln(Y) - \ln(X)| < \ln(1+\theta)\right].$$
(2.42)

Let $D = \ln(Y) - \ln(X)$ and $\delta_{\pi_0} = \ln(1+\theta_{\pi_0})$. Then $\theta_{\pi_0} = 100[\exp(\delta_{\pi_0}) - 1]\%$. This $100\theta_{\pi_0}\%$ is denoted by TDI%$_{\pi_0}$.

In the case of proportional errors, all of the above unscaled and scaled agreement indices should be computed from the log transformed data. In practice, we have encountered the proportional error case more frequently than the constant error case.

2.6 Summary of Simulation Results

A comprehensive simulation was conducted in Lin, Hedayat, Sinha, and Yang (2002) to study the small-sample properties of the transformed estimates of CCC, precision coefficient, accuracy coefficient, TDI, and CP. The results showed excellent agreement with the theoretical values from normal samples even when $n = 15$. However, these estimates are not expected to be robust against outliers or large deviations from normality or log-normality. The robustness issues of the CCC have been addressed in King and Chinchilli (2001a, 2001b), using M-estimation or using a power function of the absolute value of D to compute the CCC.

2.7   Asymptotic Power and Sample Size

In assessing agreement, the null and alternative hypotheses should be reversed. The conventional rejection region actually is the region of declaring agreement (one-sided). Asymptotic power and sample size calculation should proceed by the above principle. These powers of CCC, TDI, and CP were compared in Lin, Hedayat, Sinha, and Yang (2002). The results showed that the TDI and CP estimates have similar power, and are superior to CCC, but they are valid only under the normality assumption. Therefore, for inference, TDI and CP are superior to CCC. However, the CCC and precision and accuracy coefficients remain very useful and informative tools, as is evident from the following examples. In Chapter 4, we will discuss the sample size subject in greater detail.

2.8   Examples

2.8.1   Example 1: Methods Comparison

This example was presented in Lin, Hedayat, Sinha, and Yang (2002). DCLHb is a treatment solution containing oxygen-carrying hemoglobin. The DCLHb level in a patient's serum is routinely measured by the Sigma method. The simpler HemoCue method was modified to reproduce the DCLHb values of the Sigma method. Serum samples from 299 patients over a 50–2,000 mg/dL range were collected. The DCLHb values of each sample were measured by both methods twice, and the averages of the duplicate values were evaluated. The client required with 95% confidence that the within-sample total deviation be less than 15% of the total deviation. This means that the allowable CCC was $1 - 0.15^2 = 0.9775$. The client also needed with 95% confidence that at least 90% of the HemoCue observations be within 150 mg/dL of the targeted Sigma values. This means that the allowable $TDI_{0.9}$ was 150 mg/dL, or that the allowable CP_{150} was 0.9.

The results are presented in Fig. 2.1 and Table 2.1. The plot indicates that the within-sample error is relatively constant across the clinical range. The plot also indicates that the HemoCue accuracy is excellent and that the precision is adequate. The CCC estimate is 0.987, which means that the within-sample total deviation is about 11.6% of the total deviation. The CCC one-sided lower confidence limit is

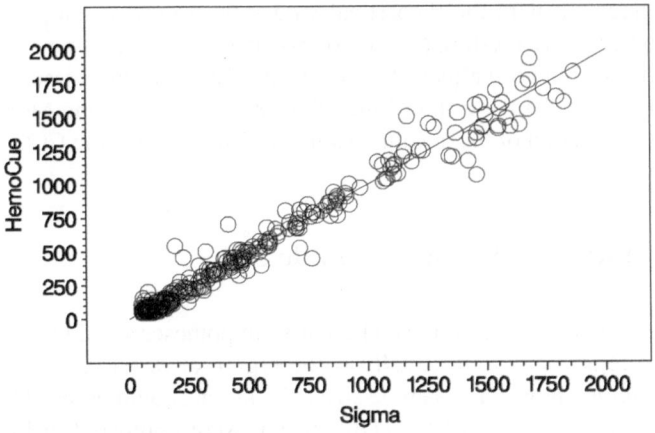

Fig. 2.1   HemoCue and Sigma readings on measuring DCLHb

Table 2.1   Agreement statistics for HemoCue and Sigma readings on measuring DCLHb

Statistics	CCC	Precision coefficient	Accuracy coefficient	$TDI_{0.9}$	CP_{150}	RBS
Estimate	0.9866	0.9867	0.9999	127.5	0.9463	0.00
95% Conf. limit	0.9838	0.9839	0.9989	136.4	0.9276	–
Allowance	0.9775	–	–	150.0	0.9000	–

"–" means "not applicable"

0.984, which is greater than 0.9775. The precision coefficient estimate is 0.987 with a one-sided lower confidence limit of 0.984. The accuracy coefficient estimate is 0.9999 with a one-sided lower confidence limit of 0.9989. The $TDI_{0.9}$ estimate is 127.5 mg/dL, which means that 90% of HemoCue observations are within 127.5 mg/dL of their target values. The one-sided upper confidence limit for $TDI_{0.9}$ is 136.4 mg/dL, which is less than 150 mg/dL. Finally, the CP_{150} estimate is 0.946, which means that 94.6% of HemoCue observations are within 150 mg/dL of their target values. The one-sided lower confidence limit for CP_{150} is 0.928, which is greater than 0.9. Therefore, the agreement between HemoCue and Sigma is acceptable with excellent accuracy and adequate precision. The relative bias squared is estimated to be near zero, indicating that the approximation of TDI should be excellent.

2.8.2   Example 2: Assay Validation

This example was presented in Lin, Hedayat, Sinha, and Yang (2002). FVIII is a clotting agent in plasma. The FVIII assay uses a marker with varying dilutions of known FVIII activities to form a standard curve. The assay started at 1:5 or 1:10

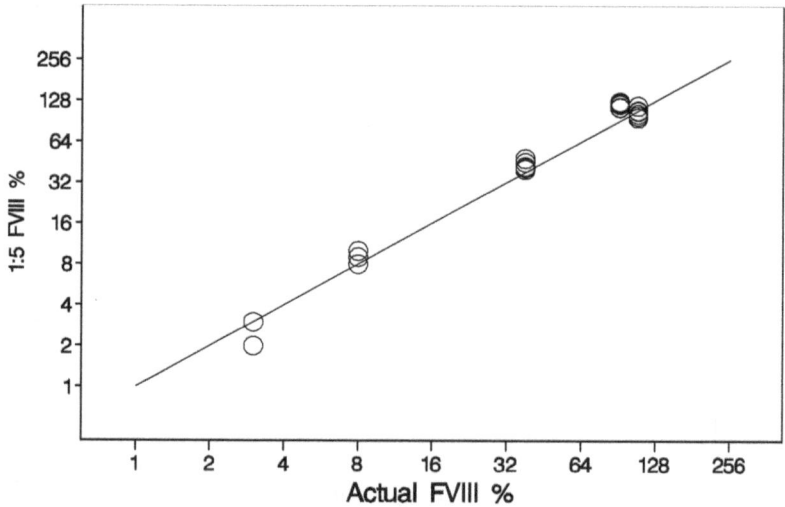

Fig. 2.2 Observed FVIII assay results versus targeted values started at 1:5

serial dilutions were prepared until they reached the target values. Target values were fixed at 3%, 8%, 38%, 91%, and 108%. Six samples were assayed per target value. The error was expected to be proportional mainly due to dilutions. The client needed with 95% confidence that the within-sample total deviation be less than 15% of the total deviation. This means that the allowable CCC was $1 - 0.15^2 = 0.9775$. The client also needed with 95% confidence that 80% of FVIII observations be within 50% of target values (note that this is the percentage of the measuring unit, which is also a percentage). This means that the allowable $TDI\%_{0.8}$ was 50%, or that the allowable $CP_{50\%}$ was 0.8.

Figures 2.2 and 2.3 present the results started at 1:5 and at 1:10 serial dilutions for these plots of observed FVIII assay results versus targeted values in $\log_2$ scale. Note that there are overlying observations in the plots. Specifically, in Fig. 2.2, four replicate readings of 3% and duplicate readings of 2% are observed at the target value of 3%, and circles at the target value of 8% represent duplicate readings of 8%, 9%, and 10%. Duplicate readings of 45% are observed at target values of 38%. Also note that in Fig. 2.3, four replicate readings of 5% and duplicate readings of 4% are observed at the target value of 3%. Three replicate readings of 11% and duplicate readings of 12% are observed at the target value of 8%. Duplicate readings of 49% are observed at target values of 38%. Duplicate readings of 124% are observed at target values of 91%. The plots indicate that the within-sample error is relatively constant across the target values in log scale. The precision is good for both assays started at 1:5 and at 1:10 serial dilutions, but the accuracy is not as good for the assay started at 1:10 serial dilutions.

Tables 2.2 and 2.3 present the agreement statistics started at 1:5 and 1:10 serial dilutions. For the assay started at 1:5 serial dilutions, the CCC is estimated to be 0.992, which means that the within-sample total deviation is about 9.1% of the

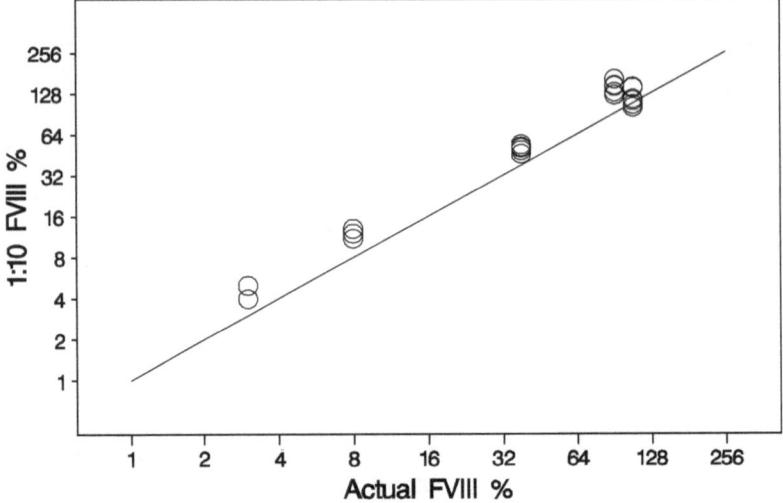

Fig. 2.3 Observed FVIII assay results versus targeted values started at 1:10

Table 2.2 FVIII assay results started at 1 : 5

Statistics	CCC	Precision coefficient	Accuracy coefficient	$TDI\%_{0.8}$	$CP_{50\%}$	RBS
Estimate	0.9917	0.9942	0.9975	27.35	0.9653	0.12
95% Conf. limit	0.9875	0.9908	0.9935	35.01	0.8921	–
Allowance	0.9775	–	–	50.0	0.8000	–

"–" means "not applicable"

Table 2.3 FVIII assay results started at 1:10

Statistics	CCC	Precision coefficient	Accuracy coefficient	$TDI\%_{0.8}$	$CP_{50\%}$	RBS
Estimate	0.9669	0.9947	0.9721	58.95	0.7016	3.75
95% Conf. limit	0.9584	0.9917	0.9638	69.01	0.5898	–
Allowance	0.9775	–	–	50.0	0.8000	–

"–" means "not applicable"

total deviation. The one-sided lower confidence limit is 0.987, which is greater than 0.9775. The precision coefficient is estimated to be 0.994 with a one-sided lower confidence limit of 0.991. The accuracy coefficient is estimated to be 0.998 with a one-sided lower confidence limit of 0.994. $TDI\%_{0.8}$ is estimated to be 27.3%, which means that 80% of observations are within 27.3% change from target values (percentage of percentage values). The one-sided upper confidence limit is 35.0%, which is less than 50%. Finally, $CP_{50\%}$ is estimated to be 0.965, which means that 96.5% of observations are within 50% change from target values. The one-sided lower confidence limit is 0.892, which is greater than 0.8. The agreement between the FVIII assay and the actual concentration is acceptable with good precision and accuracy. The relative bias squared is estimated to be 0.12, so that the approximation of TDI should be excellent.

For the assay started at 1:10 serial dilutions, the CCC is estimated to be 0.967, which means that the within-sample total deviation is about 18.2% of the total deviation. The one-sided lower confidence limit is 0.958, which is less than 0.9775. The precision coefficient is estimated to be 0.995 with a one-sided lower confidence limit of 0.992. The accuracy coefficient is estimated to be 0.972 with a one-sided lower confidence limit of 0.964. $TDI\%_{0.8}$ is estimated to be 58.9%, which means that 80% of observations are within 58.9% change from target percentage values. The one-sided upper confidence limit is 69.0%, which is greater than 50%. Finally, $CP_{50\%}$ is estimated to be 0.702, which means that 70.2% of observations are within 50% change from target values. The one-sided lower confidence limit is 0.590, which is less than 0.8. The agreement between the FVIII assay and actual concentration had good precision but is not acceptable due to mediocre accuracy. The relative bias squared is estimated to be 3.75, which is less than 8.0, so that the approximation of TDI should be acceptable.

2.8.3 Example 3: Assay Validation

This example was presented in Lin and Torbeck (1998). A study to validate an amino acid analysis test method was conducted. Solutions were prepared at approximately 90%, 100%, and 110% of label concentration of the amino acids, each containing nine determinations (observed values). Target values were determined based on their average molecular weights, which were much more precise and accurate but were still measured with random error. For each test method we compute the estimates of CP, TDI, CCC, precision coefficient, accuracy coefficient, and their confidence limits. It is debatable whether we should treat the target values as random or fixed because they were average values. We therefore take the more conservative approach by treating target values as random, which yields the same estimates of agreement statistics but with a larger respective standard error for each estimate.

The observed and target values were expressed as a percentage of label concentration. Using estimates of the CCC components, coefficients of accuracy (c_a) and precision (r), four out of 30 amino acids were chosen for illustration, each representing an example of distinctive precise and/or accurate situations. These four amino acids and their label concentrations were glycine (1 g/L), ornithine (6.4 g/L), L-threonine acid (3.6 g/L), and l-methionine (2 g/L).

The range of these data was approximately 20% (90%–110%) of label concentration. The client needed with 95% confidence that at least 80% of observations be within 3% of target values. This means that the 95% upper limit of $TDI_{0.8}$ must be less than 3, or that the 95% lower limit of CP_3 must be greater than 0.8. Note that the measurement unit is in percentage, and the error structure was assumed constant across the data range. The client did not specify a criterion for the CCC.

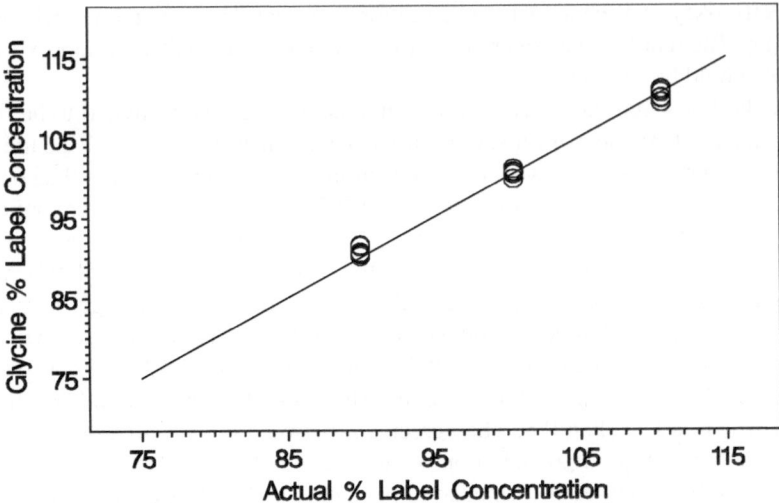

Fig. 2.4 Observed measures versus target values of glycine

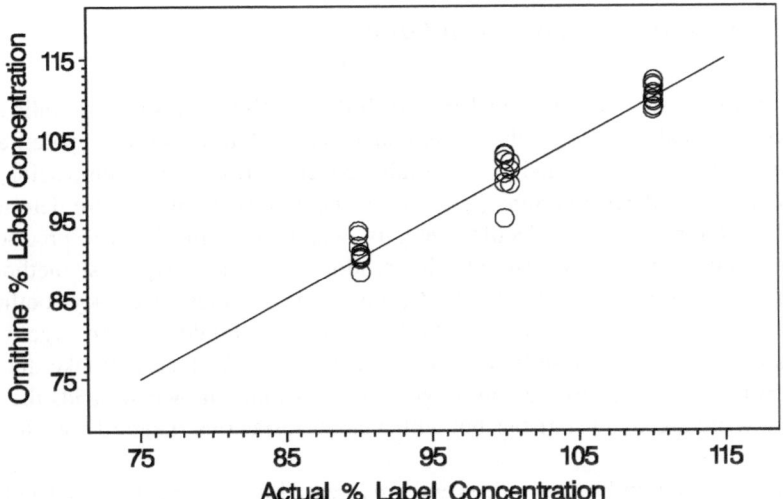

Fig. 2.5 Observed measures versus target values of ornithine

Figures 2.4–2.7 present the plots and Tables 2.4–2.7 present the agreement statistics for glycine, ornithine, L-threonine acid, and L-methionine, respectively.

The results for glycine are accurate and precise, with CCC = 0.996 (0.994), $r = 0.998$ (0.996), $c_a = 0.998$ (0.996), $\text{TDI}_{0.8} = 0.93$ (1.18), and $\text{CP}_3 = 0.9999$ (0.9966). Values presented in parentheses represent the respective 95% lower or upper confidence limit. More than 80% of observations are within 0.93 of target values. The 95% upper confidence limit of $\text{TDI}_{0.8}$ is 1.18, which is within the

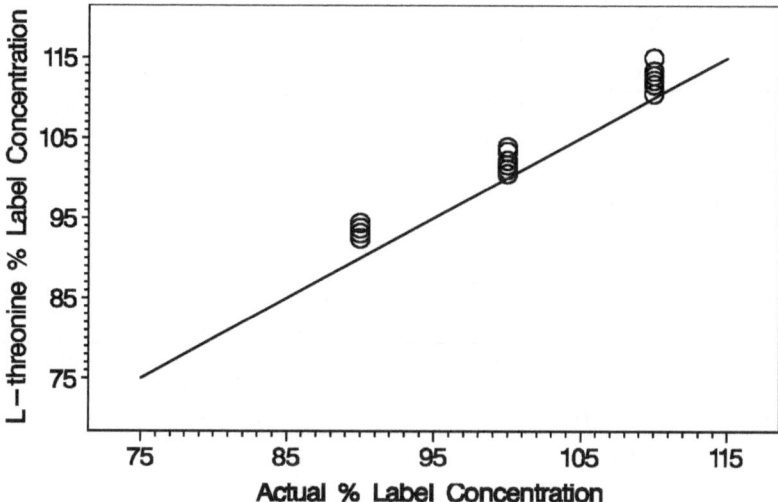

Fig. 2.6 Observed measures versus target values of L-threonine

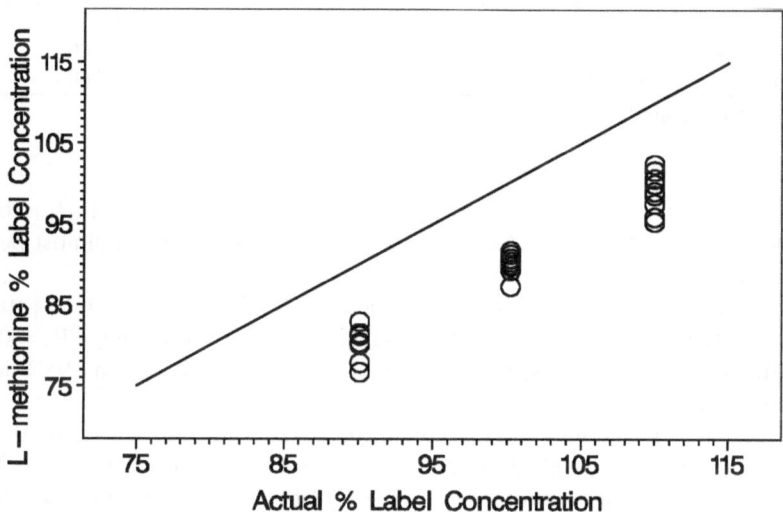

Fig. 2.7 Observed measures versus target values of L-methionine

Table 2.4 Glycine results ($n = 27$)

Statistics	CCC	Precision coefficient	Accuracy coefficient	$TDI_{0.8}$	CP_3	RBS
Estimate	0.9962	0.9981	0.9981	0.93	0.9999	0.08
95% Conf. limit	0.9937	0.9963	0.9961	1.18	0.9966	–
Allowance	0.9900	–	–	3.0	0.8000	–

"–" means "not applicable"

Table 2.5 Ornithine results ($n = 27$)

Statistics	CCC	Precision coefficient	Accuracy coefficient	$TDI_{0.8}$	CP_3	RBS
Estimate	0.9738	0.9759	0.9978	2.45	0.8703	0.08
95% Conf. limit	0.9502	0.9535	0.9801	3.09	0.7496	–
Allowance	0.9900	–	–	3.0	0.8000	–

"–" means "not applicable"

Table 2.6 L-threonine results ($n = 27$)

Statistics	CCC	Precision coefficient	Accuracy coefficient	$TDI_{0.8}$	CP_3	RBS
Estimate	0.9444	0.9905	0.9534	3.61	0.6557	4.44
95% Conf. limit	0.9084	0.9814	0.9222	4.14	0.5188	–
Allowance	0.9900	–	–	3.0	0.8000	–

"–" means "not applicable"

Table 2.7 L-methionine results ($n = 27$)

Statistics	CCC	Precision coefficient	Accuracy coefficient	$TDI_{0.8}$	CP_3	RBS
Estimate	0.5308	0.9723	0.5459	13.68	0.0001	26.11
95% Conf. limit	0.3991	0.9464	0.4280	14.89	0.0000	–
Allowance	0.9900	–	–	3.0	0.8000	–

"–" means "not applicable"

allowable 3%. The 95% lower confidence limit of CP_3 is 0.997, which is better than the allowable 0.8. The CCC estimate is near 1, indicating an almost perfect agreement.

The results of ornithine are accurate but less precise, with CCC $= 0.974$ (0.95), $r = 0.976$ (0.954), $c_a = 0.998$ (0.98), $TDI_{0.8} = 2.45$ (3.09), and $CP_3 = 0.870$ (0.750). The 95% upper confidence limit of $TDI_{0.8}$ is 3.09, and the 95% lower confidence limit of CP_3 is 0.750.

The results of L-threonine are inaccurate but precise, with CCC $= 0.944$ (0.908), $r = 0.991$ (0.981), $c_a = 0.953$ (0.922), $TDI_{0.8} = 3.61$ (4.14), and $CP_3 = 0.656$ (0.519). The 95% upper confidence limit of $TDI_{0.8}$ is 4.14, and the 95% lower confidence limit of CP_3 is 0.519.

The results of L-methionine are inaccurate and imprecise, with CCC $= 0.531$ (0.399), $r = 0.972$ (0.946), $c_a = 0.546$ (0.428), $TDI_{0.8} = 13.68$ (14.89), and $CP_3 = 0.0001$ (0.0000). The 95% upper confidence limit of $TDI_{0.8}$ is 14.89, and the 95% lower confidence limit of CP_3 is almost zero. Note that the $TDI_{0.8}$ estimate of the L-methionine assay is conservative, since the estimate of its relative bias squared value is large.

In summary, only the glycine assay in this example meets the client's criterion.

2.8.4   Example 4: Lab Performance Process Control

This example was presented in Lin (2008). For quality control of clinical laboratories, control materials of various concentrations were randomly sent to laboratories for testing. The test results were to satisfy the proficient testing (PT) criterion. The PT criterion for each lab test of the Clinical Laboratory Improvement Amendments (CLIA) Final Rule (2003) http://wwwn.cdc.gov/clia/regs/subpart_i.aspx#493. 929 required that 80% of observations be within a certain percentage or unit of the target concentration for measuring control materials. The target concentrations usually were the average of control materials across a peer group of labs using similar instruments. Such a criterion lends itself directly for using the $TDI\%_{0.8}$ or $TDI_{0.8}$.

For each of the majority of lab measurements, laboratories were required to test commercial control materials at least once a day for at least two concentrations (low and high). Daily glucose values of 116 laboratory instruments were monitored. Based on accuracy and precision indices, we selected four laboratory instruments with four distinct combinations of precision and accuracy. For each laboratory instrument we computed $TDI\%_{0.9}$, CCC, and precision and accuracy coefficients. Here, we chose to monitor 90% for a cushion, instead of 80%, of observations across all levels that were within $TDI\%_{0.9}$ or $TDI_{0.9}$ units of targets. We can translate from $TDI_{0.9}$ to $TDI_{0.8}$ by multiplying by $1.282/1.645 = 0.779$. The target values were computed as the average of these 116 laboratory instruments. For glucose, this PT criterion was 10% or 6 mg/dL, whichever was larger. The range of these data was around 70–270 mg/dL. In this case, the 10% value was the PT criterion (always larger than 6 mg/dL).

For each lab instrument, we computed the preceding agreement statistics for each calendar month and for the last available 50 days (current window). Across the 116 lab instruments, we computed the group geometric mean (GM), one standard deviation (1-SD), and two standard deviation (2-SD) upper limits with 3-month average $TDI\%_{0.9}$ values as benchmarks. Note that the distribution of $TDI\%_{0.9}$ was shown to be log-normal. Here, the confidence limit of $TDI\%_{0.9}$ per lab instrument was not computed, because we were using the population benchmarks. Therefore, it is irrelevant whether the target values were treated as random or fixed.

Figures 2.8–2.11 present the plots of the four selected cases. For each case, the left-hand plot presents the usual agreement plot of observations versus target values for the current window, and each plotted symbol (circle) represents the daily glucose value against the daily average glucose values across 116 labs. The right-hand plot monitors the quality control results over a selected time window based on $TDI\%_{0.9}$ values. We chose to monitor a rolling three-completed-month window (June, July, and August in this case) plus the current window. Each plotted symbol (dot) represents the monthly or current window $TDI\%_{0.9}$ value. Also presented are the population benchmarks of geometric mean, 1-SD, and 2-SD upper limits, and the PT criterion (PTC) of 10% (dashed line). Although the CCC, precision coefficient, and accuracy coefficient are not shown in the right-hand plot, these

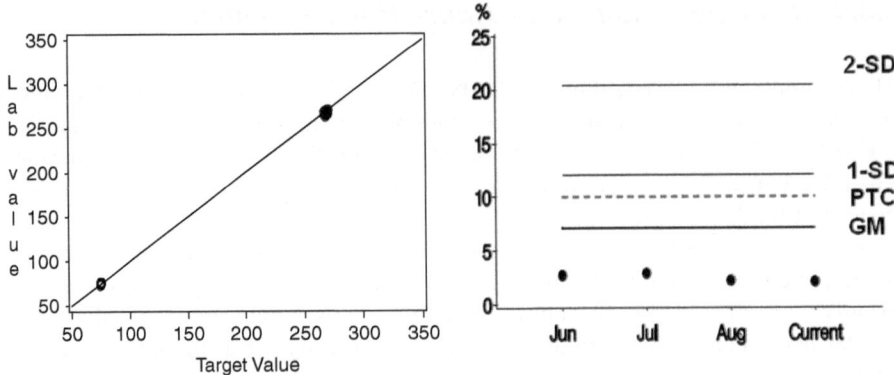

Fig. 2.8 Observed glucose measures versus target values of lab instrument for the current window and the control chart based on TDI%$_{0.9}$: almost perfect

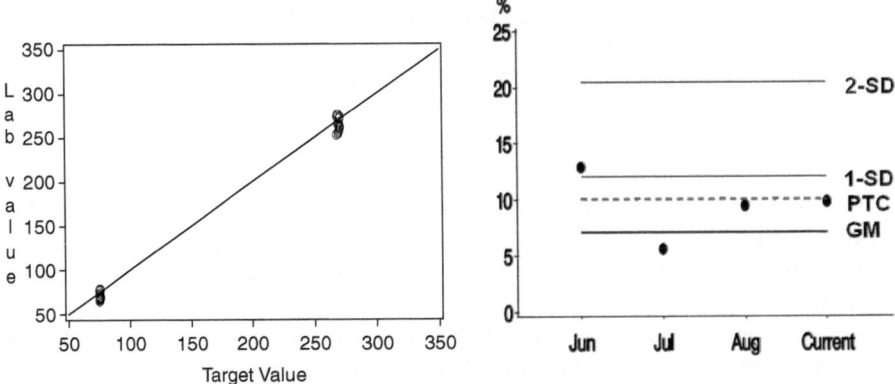

Fig. 2.9 Observed glucose measures versus target values of lab instrument for the current window and the control chart based on TDI%$_{0.9}$: imprecise but accurate

values of the current window were used to select the four instruments that are presented here. The use of CP can be helpful here. However, the CP values have difficulty discriminating among good instruments when they all have very high CP values.

Figure 2.8 shows the best-performing lab instrument among all 116 lab instruments, with CCC = 0.9998, r^2 = 0.9997, c_a = 0.9999, and TDI%$_{0.9}$ = 2.1% for the current window. It has an almost perfect CCC, and its TDI%$_{0.9}$ values are around 2%–3%.

Figure 2.9 shows a less-precise but accurate lab instrument, with CCC = 0.996, r^2 = 0.996, c_a = 0.998, and TDI%$_{0.9}$ = 9.8% for the current window. Its values rank at around 2/3 (slightly greater than 1-SD value) among its peers in June, slightly better than its peer average in July, and at around the PTC level in August and the current window.

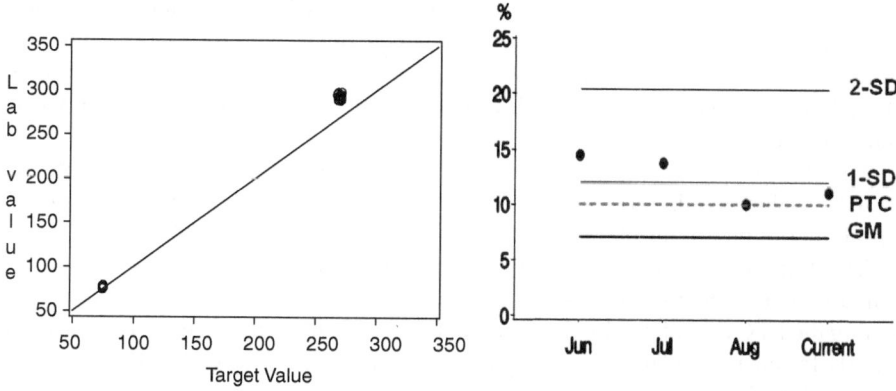

Fig. 2.10 Observed glucose measures versus target values of lab instrument for the current window and the control chart based on TDI%$_{0.9}$: precise but inaccurate

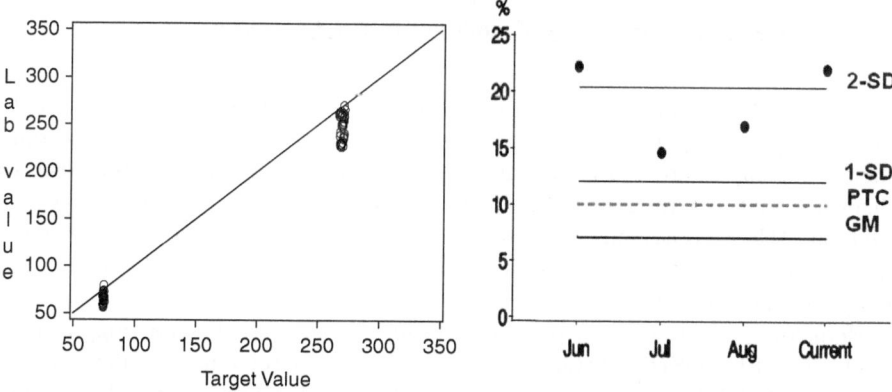

Fig. 2.11 Observed glucose measures versus target values of lab-instrument for the current window and the control chart based on TDI%$_{0.9}$: imprecise and inaccurate

Figure 2.10 shows a precise but inaccurate lab instrument, with CCC = 0.995, $r^2 = 0.9996$, $c_a = 0.995$, and TDI%$_{0.9}$ = 11% for the current window. Its TDI%$_{0.9}$ values are between 1-SD and 2-SD values of its peers in June and July, at around the PTC level in August, and slightly worse than the PTC in the current window.

Figure 2.11 shows the worst-performing lab instrument among all 116 lab instruments, with CCC = 0.983, $r^2 = 0.991$, $c_a = 0.988$, and TDI%$_{0.9}$ = 22% for the current window. Its TDI%$_{0.9}$ values are between 1-SD and 2-SD values of its peers in July and August, and worse than the 2-SD value of its peers in June and the current window.

This example conveys a few lessons. First, it is difficult to judge agreement solely by the CCC, precision coefficient, and accuracy coefficient in their absolute values. All comparisons should be judged in relative terms. Here, even the worst CCC value shown in Fig. 2.11 is 0.983. Such a high value here is due to the large study range of

70–270 mg/dL. Note that as stated earlier, the CCC, precision coefficient, accuracy coefficient, and ICC depend largely on the study range. Comparisons among any of these are valid only with similar study ranges. In this example, the study ranges (based on peer means) are identical. It is important to report the study range when reporting these statistics.

Second, not all lab instruments are created equal. Their quality, in terms of $TDI\%_{0.9}$ values, could range from 2% to 22%, which is quite diverse. It is important to submit our blood samples to a lab with a good reputation for quality.

Third, the PTC value is between the GM and 1-SD benchmarks for measuring glucose, which means that about one-third of the lab instruments are in danger of failing the PTC as dictated by the CLIA 2003 Final Rule. Perhaps the PTC is set too strictly for measuring glucose. Note that we use $TDI\%_{0.9}$ instead of $TDI\%_{0.8}$ values for a cushion here.

2.8.5   Example 5: Clinical Chemistry and Hematology Measurements That Conform to CLIA Criteria

The data for this example were obtained through the clinical labs within the research and development organization at Baxter Healthcare Corporation. Analysis of serum chemistry analytes previously validated using the Hitachi 911 (Roche Diagnostics) chemistry analyzer (reference or gold standard assays) was to be converted to the Olympus AU400e (Olympus America, Inc.) chemistry analyzer (test or new assays). Assay comparisons were performed by analyzing approximately 50–60 samples per assay on both analyzers.

Hematology analyses were performed using two automated hematology instruments: the CELL DYN 3500 (Abbott Diagnostics) as the reference and the ADVIA 2120 (SIEMENS Healthcare Diagnostics) as the test. Analyses were performed on whole blood samples drawn into tubes containing EDTA anticoagulant. A total of 93 samples from 16 humans, 18 rabbits, 19 rats, 20 dogs, and 20 pigs were each tested once on both instruments. All species were combined to establish wider data ranges, and assay performance was not expected to depend on species.

Evaluations included clinical chemistry analytes of albumin, BUN, calcium, chloride, creatinine, glucose, iron, magnesium, potassium, sodium, total protein, and triglycerides; and hematology analytes of hemoglobin, platelet count, RBC, and WBC. Any analyte without a PTC or values outside the data range was not included in the evaluations.

Table 2.8 presents the data range, PT criterion, and estimates and confidence limits (in parentheses) of agreement statistics for each of the above analytes. Data ranges of clinical chemistry analytes were acquired from the Olympus Chemistry Reagent Guide "Limitations of the Procedure" section for each assay (Olympus America, Inc. Reagent Guide Version 3.0. May 2007. Irving, TX). Data ranges of hematology analytes were acquired from ADVIA 2120 Version 5.3.1–MS2005 Bayer (now Siemens) Healthcare LLC.

Table 2.8 Agreement statistics against the PT criteria (PTC) for clinical chemistry and hematology analytes

Analyte	Range	PTC	n	CCC	Precision Coef.	Accuracy Coef.	$TDI_{0.8}$[a]	CP_{PTC}[b]	RBS[c]
Albumin (g/dL)	**1.5–6.0**	**10%**	**58**	**0.723 (0.626)**	**0.859 (0.789)**	**0.842 (0.768)**	**23.6% (27.29%)**	**0.379 (0.309)**	**1.19**
BUN ≤ 22 (mg/dL)	2–22	2	44	0.993 (0.989)	0.996 (0.993)	0.997 (0.994)	0.77 (0.91)	1.000 (0.996)	0.53
BUN >22 (mg/dL)	22–130	9%	16	0.999 (0.998)	0.999 (0.998)	1.000 (0.999)	3.26% (4.48%)	0.999 (0.954)	0.05
Calcium (mg/dL)	**4–18**	**1**	**60**	**0.873 (0.833)**	**0.996 (0.994)**	**0.876 (0.838)**	**1.58 (1.69)**	**0.302 (0.223)**	**11.7**
Chloride (mmol/L)	50–200	5%	61	0.991 (0.987)	0.996 (0.993)	0.995 (0.992)	2.44% (2.8%)	0.995 (0.982)	0.81
Creatinine Enz <2 (mg/dL)	0.2–2	0.3	50	0.989 (0.982)	0.990 (0.983)	0.999 (0.996)	0.08 (0.09)	1.000 (1.000)	0.06
Creatinine Jaffe <2 (mg/dL)	0.2–2	0.3	47	0.963 (0.942)	0.981 (0.969)	0.981 (0.966)	0.14 (0.16)	0.998 (0.989)	0.92
Glucose >60 (mg/dL)	60–800	10%	55	0.999 (0.998)	0.999 (0.998)	1.000 (0.999)	4.28% (5.01%)	0.997 (0.986)	0.32
Iron (ug/dL)	10–1000	20%	59	0.994 (0.991)	0.997 (0.995)	0.997 (0.995)	9.74% (11.45%)	0.987 (0.961)	0.03
Magnesium (mg/dL)	0.5–8	25%	61	0.970 (0.958)	0.996 (0.993)	0.974 (0.963)	14.41% (15.71%)	0.999 (0.995)	5.82
Potassium (mmol/L)	1–10	0.5	59	0.996 (0.995)	0.998 (0.997)	0.998 (0.997)	0.14 (0.16)	1.000 (1.000)	0.43
Sodium (mmol/L)	50–200	4	59	0.994 (0.991)	0.995 (0.993)	0.999 (0.997)	1.77 (2.06)	0.996 (0.984)	0.16
Total Protein (g/dL)	3–12	10%	56	0.993 (0.989)	0.993 (0.989)	1.000 (0.997)	2.31% (2.71%)	1.000 (1.000)	0.03
Triglycerides (g/dL)	10–1000	25%	56	0.997 (0.996)	0.998 (0.998)	0.999 (0.998)	8.2% (9.56%)	1.000 (0.999)	0.59
Hemoglobin (g/dL)	1–22.5	7%	93	0.967 (0.957)	0.997 (0.996)	0.969 (0.961)	6.02% (6.4%)	0.942 (0.903)	7.08
Platelet Count (×103/ˉL)	**10–3500**	**25%**	**84**	**0.917 (0.884)**	**0.926 (0.896)**	**0.990 (0.973)**	**31.72% (36.79%)**	**0.695 (0.629)**	**0.04**
RBC (×106/μL)	0.1–12	6%	92	0.988 (0.984)	0.997 (0.995)	0.992 (0.989)	4.13% (4.52%)	0.966 (0.937)	2.22
WBC (×103/ˉL)	**0.1–100**	**15%**	**93**	**0.942 (0.922)**	**0.958 (0.941)**	**0.984 (0.972)**	**21.34% (24.41%)**	**0.640 (0.576)**	**0.08**

Note: Shown in parentheses is the 95% upper (TDI) or lower (CCC, precision, accuracy, CP) confidence limit. Boldface analytes are those that failed the PTC.
[a] Total deviation index to cover 80% of absolute difference or % change. The 95% upper limit should be less than PTC or PTC%.
[b] Coverage probability of values within the PTC. The 95% lower limit should be greater than 0.8.
[c] The relative bias squared (RBS) must be less than 8 in order for the approximate TDI to be valid. Otherwise, the TDI estimate is conservative depending on the RBS value.

For evaluation of BUN, the PTC is 9% for values greater than or equal to 22 mg/dL or 2 mg/dL for values less than or equal to 22 mg/dL, since 9% of 22 is approximately 2. The criterion used for creatinine Enz and Jaffe is 0.3 mg/dL, since only values less than 2 mg/dL were evaluated. Glucose was evaluated for values greater than 60 mg/dL with the criterion 10%.

Figure 2.12 presents the agreement plots of the above analytes. The analytes of albumin, calcium, platelet count, and WBC had precision and/or accuracy problems, while the other analytes appeared to perform well. Comparing the 95% upper confidence limit of TDI against the PTC or PTC% or the 95% lower confidence limit CP against 0.8, all but the analytes of albumin, calcium, platelet count, and WBC (shown boldface in Table 2.8) pass the PTC with 95% confidence.

Table 2.9 presents the results of traditional statistical analyses based on paired t-test and ordinary regression. The results from Deming (orthogonal) regression by treating X as a random variable are not shown because they are more or less similar to those of ordinary regression. Table 2.9 shows the data range, sample size, paired t-test p-value, intercepts, slopes, and the testing of intercept (0) and slope (1) of ordinary regressions.

The paired t-test rejects the agreement of extremely well performing analytes, that is, BUN $\leq$ 22 mg/dL, chloride, creatinine Jaffe, glucose, magnesium, potassium, sodium, triglycerides, hemoglobin, and RBC, with $p = 0.003$ for sodium and $p < 0.001$ for the others. These rejections correspond to the left-hand plot of Fig. 1.2 and are due primarily to near-zero residual variance and/or large sample size. On the other hand, the paired t-test accepts ($p = 0.062$) the agreement of platelet count. Such failure to reject corresponds to the right-hand plot of Fig. 1.2, due primarily to its large residual variance.

In terms of ordinary regression, tests of intercept (0) and/or slope (1) (LS01) reject ($p < 0.05$) the agreement of the extremely well performing analytes of BUN $\leq$ 22 mg/dL, chloride, creatinine Jaffe, iron, magnesium, potassium, sodium, triglycerides, and hemoglobin. These rejections are similar, although not identical, to the paired t-test for the same reasons. On the other hand, LS01 accepts ($p \geq 0.373$) the agreement of platelet count for the same reasons as paired t-test.

In terms of CCC, precision and accuracy for clinical chemistry analysts of BUN, chloride, glucose, iron, magnesium, potassium, sodium, total protein, triglycerides have excellent agreement (CCC > 0.99) between measurements from the Olympus AU400e and Hitachi 911 instruments. Both have excellent precision (>0.99) and accuracy (>0.99).

Creatinine Enz and creatinine Jaffe comfortably pass the PTC. Because their data ranges are small, from 0.2 to 1.8 mg/dL, their CCC and precision and accuracy coefficients become relatively lower. Magnesium also pass the PTC by a comfortable margin. It has excellent precision (0.9956) but relatively smaller accuracy (0.9738), because most Olympus values are smaller than Hitachi values by a negligible amount.

For clinical chemistry analytes of albumin and calcium, the lab has difficulties proving the equivalence of measurements from the Olympus AU400e and Hitachi 911 instruments. Albumin measurements from both the Olympus AU400e and Hitachi 911 are neither accurate (0.8422) nor precise (0.8589). We are 95%

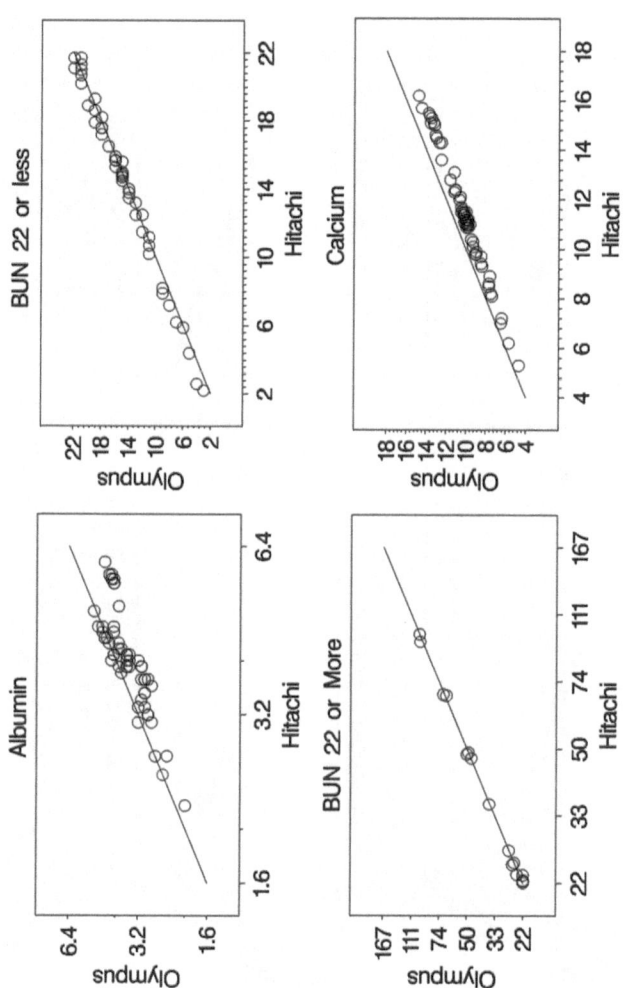

Fig. 2.12 Agreement plots

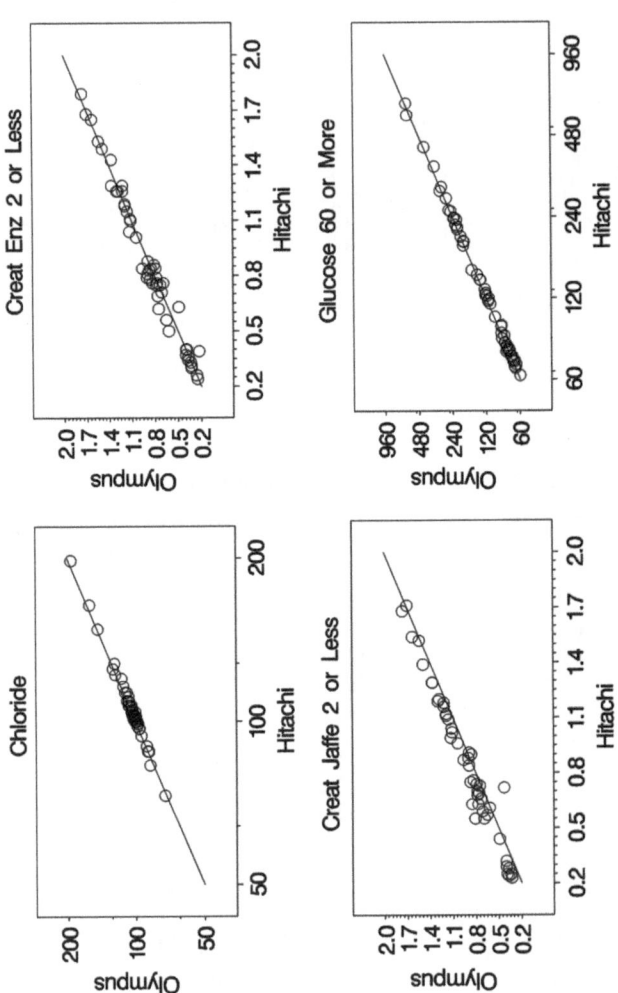

Fig. 2.12 (continued)

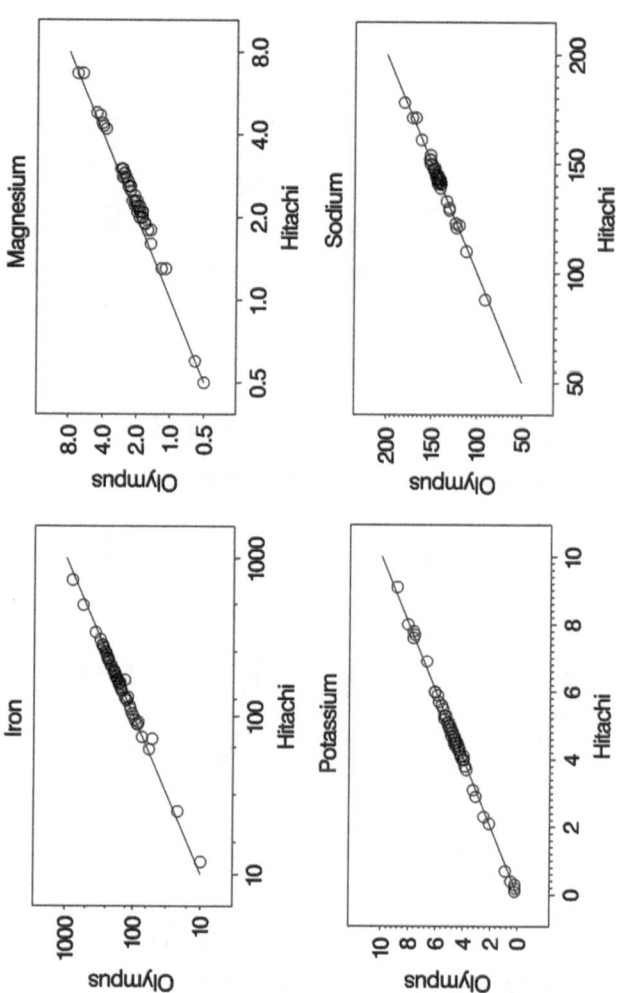

Fig. 2.12 (continued)

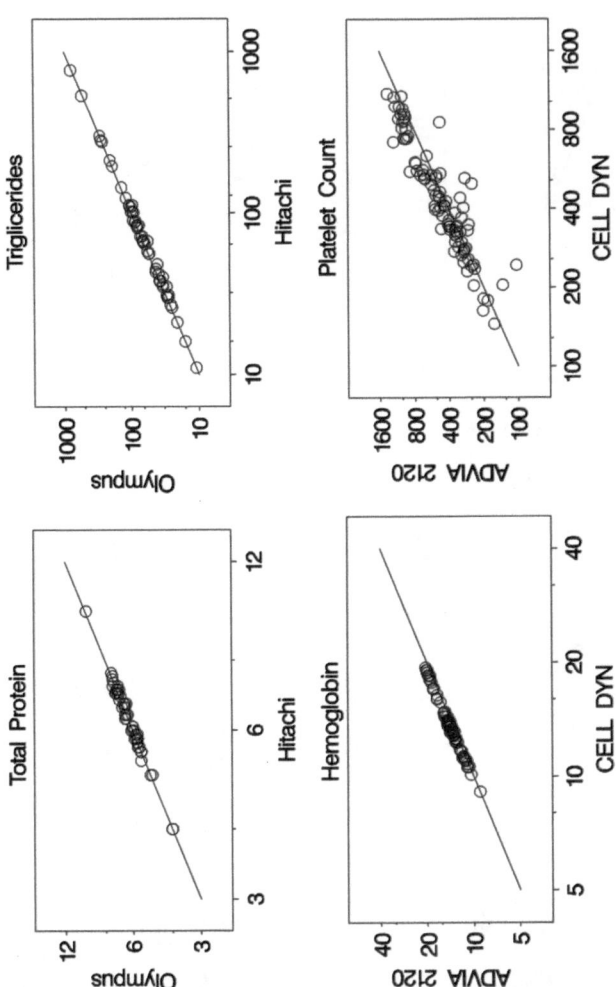

Fig. 2.12 (continued)

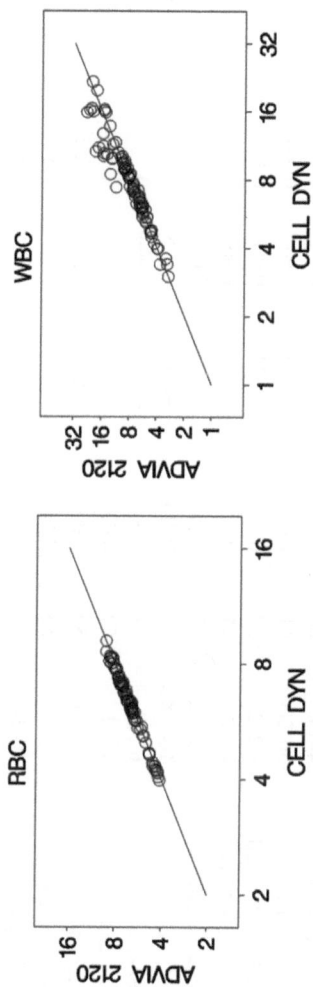

Fig. 2.12 (continued)

Table 2.9 Traditional statistics for clinical chemistry and hematology analytes

Analyte	Range	n	Paired t-test p-value	Intercept	Slope	Intercept = 0 p-value	Slope = 1 p-value
Albumin (g/dL)	1.5–6.0	58	<0.001	0.181	0.783	0.045	0.001
BUN ≤ 22 (mg/dL)	2–22	44	<0.001	0.848	0.966	0	0.014
BUN >22 (mg/dL)	22–130	16	0.365	−0.037	1.012	0.401	0.329
Calcium (mg/dL)	4–18	60	<0.001	0.15	0.884	0.205	<0.001
Chloride (mmol/L)	50–200	61	<0.001	0.16	0.963	0.005	0.002
Creatinine Enz <2 (mg/dL)	0.2–2	50	0.093	0.026	0.986	0.196	0.517
Creatinine Jaffe <2 (mg/dL)	0.2–2	47	<0.001	0.107	0.962	0	0.186
Glucose >60 (mg/dL)	60–800	55	<0.001	−0.049	1.007	0.129	0.306
Iron (ug/dL)	10–1000	59	0.214	−0.383	1.078	0	<0.001
Magnesium (mg/dL)	0.5–8	61	<0.001	−0.095	0.999	0	0.907
Potassium (mmol/L)	1–10	59	<0.001	0.22	0.967	0	<0.001
Sodium (mmol/L)	50–200	59	0.003	4.025	0.975	0.032	0.06
Total Protein (g/dL)	3–12	56	0.198	0.015	0.99	0.602	0.531
Triglycerides (g/dL)	10–1000	56	<0.001	−0.104	1.016	0.003	0.049
Hemoglobin (g/dL)	1–22.5	93	<0.001	−0.078	1.046	0.001	<0.001
Platelet Count (×103/µL)	10–3500	84	0.062	−0.211	1.042	0.461	0.373
RBC (×106/µL)	0.1–12	92	<0.001	0.005	1.012	0.776	0.194
WBC (×103/µL)	0.1–100	93	0.006	−0.196	1.116	0.009	0.001

confident that albumin measurements can deviate 27.3% (>10%) from their target values, and that the measured values conformed to the PTC only 30.9% (<80%) of the time. Calcium measurements are precise (0.9963) but not accurate (0.8759). We are 95% confident that calcium measurements can deviate 1.58 mg/dL (>1 mg/dL) from their target values, and that the measured values conform to the PTC only 22.8% (<80%) of the time.

For hemoglobin, the lab has good agreement (CCC = 0.9665) between measurements from the Advia 2120 and Cell DYN 3500 instruments, with excellent precision (0.9970) and good accuracy (0.9694). There is a small bias showing that measurements from the Advia 2120, with only one exception, are consistently higher than from the Cell DYN 3500. Note that data were not collected over the full analytical range of 5 to 22 g/dL for this analyte.

RBC counts have very good agreement (CCC = 0.9883) between measurements from the Advia 2120 and Cell DYN 3500 instruments. Both have excellent precision (0.9965) and accuracy (0.9917). There is a small bias showing that all but one of the measurements from Advia 2120 are consistently higher than those of Cell DYN 3500. Note that data are not collected over the full analytical range of 2 to $10 \times 106/\mu L$ for this analyte.

Platelet count from the Advia 2120 and Cell DYN 3500 are accurate (0.9897) but imprecise (0.9263). We are 95% confident that platelet count measurements can deviate 36.8% (>25%) from their target values and that they conformed to the PTC only 62.9% (<80%) of the time. Additionally, WBC measurements are also relatively accurate (0.9839) but imprecise (0.9576), especially for readings greater than $8 \times 103/\mu L$. The lab fails to show that these two analytes meet the PTC. We are 95% confident that WBC measurements can deviate 24.4% (>15%) from their target values, and that they conform to the PTC only 58.0% (<80%) of the time.

In summary, using the agreement statistics presented in this example, 14 out of 18 method comparison cases meet the CLIA criteria with 95% confidence. Of the four that did not meet the CLIA criteria, one is acceptable by the traditional paired t-test and regression analysis. Of the 14 that meet the CLIA criteria, 11 are rejected by the traditional paired t-test or regression analysis.

2.9   Proofs of Asymptotical Normality When Target Values Are Random

2.9.1   CCC and Precision Estimates

This proof can be seen in Lin (1989). The Z transformation of the CCC estimate can be expressed as $Z = g(m)$, where

$$\boldsymbol{m} = (m_1, m_2, m_3, m_4, m_5)'$$

$$= \left(\overline{Y}, \overline{X}, \frac{1}{n} \sum_{i=1}^{n} X_i^2, \frac{1}{n} \sum_{i=1}^{n} Y_i^2, \frac{1}{n} \sum_{i=1}^{n} Y_i X_i \right)', \tag{2.43}$$

and

$$Z = g(\boldsymbol{m}) = \frac{1}{2} \ln \left[1 + \frac{4(m_5 - m_1 m_2)}{m_3 + m_4 - 2m_5} \right].$$

The vector $\boldsymbol{m}$ is expressed as a function of sample moments, and has an asymptotic 5-variate normality with mean

$$\Theta = (\mu_y, \mu_x, \sigma_y^2 + \mu_y^2, \sigma_x^2 + \mu_x^2, \sigma_{yx} + \mu_y \mu_x)',$$

and variance $\frac{1}{n}\boldsymbol{\Sigma}$, where

$$\boldsymbol{\Sigma} = \{\lambda_{ij}\}_{5 \times 5}. \tag{2.44}$$

Here, with the assumption that $\{(Y_i, X_i) \mid i = 1, 2, \ldots, n\}$ are random samples from a bivariate normal distribution, we have

$$\lambda_{11} = \sigma_y^2,$$
$$\lambda_{12} = \lambda_{21} = \sigma_{yx},$$
$$\lambda_{22} = \sigma_x^2,$$
$$\lambda_{13} = \lambda_{31} = 2\mu_y \sigma_y^2,$$
$$\lambda_{23} = \lambda_{32} = 2\mu_y \sigma_{yx},$$
$$\lambda_{33} = 2\sigma_y^4 + 4\sigma_y^2 \mu_y^2,$$
$$\lambda_{14} = \lambda_{41} = 2\mu_x \sigma_{yx},$$
$$\lambda_{24} = \lambda_{42} = 2\mu_x \sigma_x^2,$$
$$\lambda_{34} = \lambda_{43} = 2\sigma_{yx}^2 + 4\mu_y \mu_x \sigma_{yx},$$
$$\lambda_{44} = 2\sigma_x^4 + 4\sigma_x^2 \mu_x^2,$$
$$\lambda_{15} = \lambda_{51} = \mu_x \sigma_y^2 + \mu_y \sigma_{yx},$$
$$\lambda_{25} = \lambda_{52} = \mu_y \sigma_x^2 + \mu_x \sigma_{yx},$$
$$\lambda_{35} = \lambda_{53} = 2\sigma_{yx} \mu_y^2 + 2\sigma_{yx} \sigma_y^2 + 2\mu_y \mu_x \sigma_y^2,$$
$$\lambda_{45} = \lambda_{54} = 2\sigma_{yx} \mu_x^2 + 2\sigma_{yx} \sigma_x^2 + 2\mu_y \mu_x \sigma_x^2,$$

and

$$\lambda_{55} = \sigma_y^2 \sigma_x^2 + \mu_y^2 \sigma_x^2 + \mu_x^2 \sigma_y^2 + \sigma_{yx}^2 + 2\mu_y \mu_x \sigma_{yx}.$$

It follows from the delta method or from the theory of functions of asymptotically normal vectors (Serfling 1980, Corollary 3.3) that Z is asymptotically normal with mean $\frac{1}{2} \ln \frac{1+\rho_c}{1-\rho_c}$ and variance $\frac{1}{n} \boldsymbol{d}' \boldsymbol{\Sigma} \boldsymbol{d}$, where

$$\boldsymbol{d} = (d_1, d_2, d_3, d_4, d_5)'$$

$$= \left(\left. \frac{\partial g(\boldsymbol{m})}{\partial m_1} \right|_{\boldsymbol{m}=\Theta}, \ldots, \left. \frac{\partial g(\boldsymbol{m})}{\partial m_5} \right|_{\boldsymbol{m}=\Theta} \right)'.$$

The elements of d are

$$d_1 = \frac{-\mu_x}{\sigma_y^2 + \sigma_x^2 + 2\sigma_{yx} + (\mu_y - \mu_x)^2},$$

$$d_2 = \frac{-\mu_y}{\sigma_y^2 + \sigma_x^2 + 2\sigma_{yx} + (\mu_y - \mu_x)^2},$$

$$d_3 = d_4 = \frac{-\sigma_{yx}}{\left[\sigma_y^2 + \sigma_x^2 + 2\sigma_{yx} + (\mu_y - \mu_x)^2\right]\left[\sigma_y^2 + \sigma_x^2 - 2\sigma_{yx} + (\mu_y - \mu_x)^2\right]},$$

and

$$d_5 = \frac{(\sigma_y^2 + \sigma_x^2) + (\mu_y - \mu_x)^2}{\left[\sigma_y^2 + \sigma_x^2 + 2\sigma_{yx} + (\mu_y - \mu_x)^2\right]\left[\sigma_y^2 + \sigma_x^2 - 2\sigma_{yx} + (\mu_y - \mu_x)^2\right]}.$$

After straightforward, albeit tedious algebraic calculations, it can be shown that the variance of Z is

$$\sigma_Z^2 = \frac{1}{n}d'\Sigma d$$

$$= \frac{1}{n}\left[\frac{(1-\rho^2)\rho_c^2}{(1-\rho_c^2)\rho^2} + \frac{2\rho_c^3(1-\rho_c)\upsilon^2}{\rho(1-\rho_c^2)^2} - \frac{\rho_c^4\upsilon^4}{2\rho^2(1-\rho_c^2)^2}\right]. \qquad (2.45)$$

The Z-transformed CCC estimate can approach normality much more rapidly as confirmed by the Monte Carlo experiment in Lin (1989). When $\rho_c = \rho$ and $\upsilon = 0$, (2.45) degenerates into $\frac{1}{n}$, which is the variance of the Z transformation of the precision estimate.

2.9.2   MSD Estimate

From (2.43), we can write the natural log transformation of the MSD estimate, or $W = \ln(e^2)$ as

$$W = g(m_3, m_4, m_5) = \ln(m_3 + m_4 - 2m_5).$$

By the delta method, W is asymptotically normal with mean $\ln(\epsilon^2)$ and variance $\frac{1}{n}d'\Sigma d$, where

$$\Sigma = \begin{pmatrix} \lambda_{33} & \lambda_{34} & \lambda_{35} \\ & \lambda_{44} & \lambda_{45} \\ & & \lambda_{55} \end{pmatrix}$$

and

$$\underline{d} = \left(\frac{\partial g(m)}{\partial m_3}\bigg|_{m=\Theta}, \frac{\partial g(m)}{\partial m_4}\bigg|_{m=\Theta}, \frac{\partial g(m)}{\partial m_5}\bigg|_{m=\Theta} \right)' = \left(\frac{1}{\varepsilon^2}, \frac{1}{\varepsilon^2}, -\frac{2}{\varepsilon^2} \right)'.$$

After straightforward algebraic calculations, we can show that the variance of W is

$$\sigma_W^2 = \frac{2}{n}\left[1 - \frac{(\mu_y - \mu_x)^4}{\varepsilon^4} \right]. \tag{2.46}$$

2.9.3 CP Estimate

This proof can be seen in Lin, Hedayat, Sinha, and Yang (2002). Here, we use a different approach to demonstrating the delta method. We can use a first-order approximation to compute the mean and variance of the CP estimate p_{δ_0} by

$$
\begin{aligned}
p_{\delta_0} = {}& \Phi\left(\frac{\delta_0 - \mu_d}{\sigma_d} \right) - \frac{1}{\sigma_d}\phi\left(\frac{\delta_0 - \mu_d}{\sigma_d} \right)(\overline{X}_d - \mu_d) \\
& -\frac{\delta_0 - \mu_d}{\sigma_d^2}\phi\left(\frac{\delta_0 - \mu_d}{\sigma_d} \right)(s_d - \sigma_d) - \Phi\left(\frac{-\delta_0 - \mu_d}{\sigma_d} \right) \\
& +\frac{1}{\sigma_d}\phi\left(\frac{-\delta_0 - \mu_d}{\sigma_d} \right)(\hat{\mu}_d - \mu_d) - \frac{\delta_0 + \mu_d}{\sigma_d^2}\phi\left(\frac{-\delta_0 - \mu_d}{\sigma_d} \right)(s_d - \sigma_d) \\
& +O\left[(\overline{X}_d - \mu_d)^2 \right] + O\left[(s_d - \sigma_d)^2 \right] + O\left[(\overline{X}_d - \mu_d)(s_d - \sigma_d) \right], \quad (2.47)
\end{aligned}
$$

where $\phi(x)$ is the density function of the standard normal distribution, $\overline{X}_d = \overline{Y} - \overline{X}$, and $\lim_{x \to 0} \frac{O(x)}{x} < \infty$.

Therefore, the expected value of p_{δ_0} is

$$E(p_{\delta_0}) = \pi_{\delta_0} + O\left(\frac{1}{n} \right),$$

and the asymptotic variance of p_{δ_0} becomes

$$
\begin{aligned}
\sigma_{p_{\delta_0}}^2 = {}& \frac{1}{n}\left\{ \left[\phi\left(\frac{-\delta_0 - \mu_d}{\sigma_d} \right) - \phi\left(\frac{\delta_0 - \mu_d}{\sigma_d} \right) \right]^2 \right. \\
& \left. + \frac{1}{2}\left[\frac{\delta_0 - \mu_d}{\sigma_d}\phi\left(\frac{\delta_0 - \mu_d}{\sigma_d} \right) + \frac{\delta_0 + \mu_d}{\sigma_d}\phi\left(\frac{-\delta_0 - \mu_d}{\sigma_d} \right) \right]^2 \right\} \\
& +O\left(\frac{1}{n^2} \right). \qquad\qquad\qquad\qquad\qquad\qquad\qquad\qquad\qquad\qquad (2.48)
\end{aligned}
$$

Because CP is bounded by 0 and 1, it is better to use the logit transformation for statistical inference. Let $T = \ln\left(\frac{p_{\delta_0}}{1-p_{\delta_0}}\right)$. Then the asymptotic mean of T is $\tau = \ln\left(\frac{\pi_{\delta_0}}{1-\pi_{\delta_0}}\right)$, and the asymptotic variance is $\sigma_T^2 = \frac{\sigma_{p_{\delta_0}}^2}{\pi_{\delta_0}^2(1-\pi_{\delta_0})^2}$.

2.9.4 Accuracy Estimate

This proof can be seen in Robieson (1999). This estimate, c_a, does not involve m_5 in (2.43). We redefine the new set of m vectors that are mostly uncorrelated as

$$m = (m_1, m_2, m_3, m_4)' = (\overline{Y}, \overline{X}, s_y^2, s_x^2)',$$

which has an asymptotic 4-variate normality with mean $E(m) = (\mu_y, \mu_x, \sigma_y^2, \sigma_x^2)'$ and variance $\frac{1}{n}\Sigma$, where $\Sigma = \{\lambda_{ij}\}_{4\times4}$.

Here,

$$\lambda_{11} = \sigma_y^2,$$
$$\lambda_{12} = \lambda_{21} = \sigma_{yx},$$
$$\lambda_{13} = \lambda_{14} = \lambda_{23} = \lambda_{24} = \lambda_{31} = \lambda_{41} = \lambda_{32} = \lambda_{42} = 0,$$
$$\lambda_{33} = 2\sigma_y^4,$$
$$\lambda_{34} = \lambda_{43} = 2\sigma_{yx}^2,$$

and

$$\lambda_{44} = 2\sigma_x^4.$$

The logit transformation of the accuracy estimate can be written as

$$L = g(m_1, m_2, m_3, m_4)$$
$$= \ln\left[\frac{2\sqrt{m_3 m_4}}{m_3 + m_4 + (m_1 - m_2)^2 - 2\sqrt{m_3 m_4}}\right],$$

and is asymptotically normal with mean $\ln\frac{\chi_a}{1-\chi_a}$ and variance $\frac{1}{n}d'\Sigma d$, where

$$d = (d_1, d_2, d_3, d_4)' = \left(\left.\frac{\partial g(m)}{\partial m_1}\right|_{m=\Theta}, \ldots, \left.\frac{\partial g(m)}{\partial m_4}\right|_{m=\Theta}\right)'.$$

The elements of d are

$$d_1 = -d_2 = \frac{(\mu_y - \mu_x)}{\sigma_y \sigma_x (1 - \chi_a)^2},$$

$$d_3 = \frac{1}{2\sigma_y^2 \chi_a (1 - \chi_a)^2} - \frac{1}{2\sigma_y \sigma_x (1 - \chi_a)^2},$$

and

$$d_4 = \frac{1}{2\sigma_x^2 \chi_a (1 - \chi_a)^2} - \frac{1}{2\sigma_y \sigma_x (1 - \chi_a)^2}.$$

It can be shown that the variance of L is

$$\sigma_L^2 = \frac{1}{n} d' \Sigma d$$

$$= \frac{1}{n(1 - \chi_a)^2} \left[\chi_a^2 \upsilon^2 \left(\varpi + \frac{1}{\varpi} - 2\rho \right) + \frac{1}{2} \chi_a^2 \left(\varpi^2 + \frac{1}{\varpi^2} + 2\rho^2 \right) \right.$$

$$\left. + (1 + \rho^2) (\chi_a \upsilon^2 - 1) \right]. \tag{2.49}$$

2.10 Proofs of Asymptotical Normality When Target Values Are Fixed

None of the proofs of asymptotic normality of the estimates of agreement indices other than CP with fixed target values have been given in the literature. The proofs here are much simplified compared to those in Section 2.9, because we are dealing with moments associated with Y only.

2.10.1 CCC and Precision Estimates

The $Z_{|X}$ transformation of the CCC estimate can be expressed as $Z_{|X} = g(m_{|X})$, where

$$m_{|X} = (m_{|X,1}, m_{|X,2}, m_{|X,3})' = (\overline{Y}, b_1, s_e^2)' \tag{2.50}$$

and

$$Z_{|X} = g(m_{|X}) = \frac{1}{2} \ln \left[\frac{m_{|X,3} + m_{|X,2}^2 s_x^2 + s_x^2 + (m_{|X,1} - \overline{X})^2 + 2m_{|X,2} s_x^2}{m_{|X,3} + m_{|X,2}^2 s_x^2 + s_x^2 + (m_{|X,1} - \overline{X})^2 - 2m_{|X,2} s_x^2} \right].$$

Here $Z_{|X}$ is asymptotically normal with mean $\frac{1}{2} \ln \frac{1+\rho_{c|X}}{1-\rho_{c|X}}$ and variance $\frac{1}{n} d'_{|X} \Sigma_{|X} d_{|X}$, where

$$\Sigma_{|X} = \begin{pmatrix} \sigma_y^2(1-\rho^2) & 0 & 0 \\ 0 & \sigma_y^2(1-\rho^2)/s_x^2 & 0 \\ 0 & 0 & 2\sigma_y^4(1-\rho^2)^2 \end{pmatrix}. \tag{2.51}$$

The elements of $d_{|X}$ are

$$d_{|X,1} = \frac{\rho_c^2(\mu_y - \overline{X})}{(1-\rho_c^2)^2 \rho \sigma_y s_x},$$

$$d_{|X,2} = -\frac{\rho_c}{(1-\rho_c^2)^2} \left(1 - \frac{1}{\rho_c \rho \upsilon}\right),$$

and

$$d_{|X,3} = \frac{\rho_c^2}{2(1-\rho_c^2)^2 \rho \sigma_y s_x}.$$

It can be shown that the variance of $Z_{|X}$ is

$$\sigma_{Z_{|X}}^2 = \frac{\rho_c^2(1-\rho^2)}{n(1-\rho_c^2)^2 \rho^2} \left[\varpi \upsilon^2 \rho_c^2 + (1-\rho_c\rho\varpi)^2 + \frac{\varpi^2 \rho_c^2(1-\rho^2)}{2}\right]. \tag{2.52}$$

When $\rho_c = \rho$, $\upsilon = 0$, and $\varpi = 1$, (2.52) degenerates to $\frac{1}{n}\left(1 - \frac{\rho^2}{2}\right)$, which is the variance of the Z transformation of the precision estimate.

2.10.2   MSD Estimate

From (2.50) we can write the natural logarithm of the MSD estimate, or $W_{|X} = \ln(e_{|X}^2)$, as

$$W_{|X} = g(m_{|X,1}, m_{|X,2}, m_{|X,3})$$
$$= \ln\left[(m_{|X,1} - \overline{X})^2 + m_{|X,3} + s_x^2(1 - m_{|X,2})^2\right].$$

By the delta method, $W_{|X}$ is asymptotically normal with mean $\ln(\varepsilon_{|X}^2)$ and variance $\frac{1}{n} d'_{|X} \Sigma_{|X} d_{|X}$, where $\Sigma_{|X}$ was shown in (2.51), and

$$d_{|X} = \left(\frac{2(\mu_y - \overline{X})}{\varepsilon_{|X}^2}, \frac{-2(1-\beta_1)s_x^2}{\varepsilon_{|X}^2}, \frac{1}{\varepsilon_{|X}^2}\right)'.$$

It can be shown that the variance of $W_{|X}$ is

$$
\sigma_{W_{|X}}^2 = \frac{2}{n} \left[1 - \frac{\left(\varepsilon_{|X}^2 - \sigma_e^2 \right)^2}{\varepsilon_{|X}^4} \right].
$$

2.10.3   CP Estimate

The proof can be seen in Lin, Hedayat, Sinha, and Yang (2002). Recall that in the regression model when target values are fixed, we assumed that e_Y has a normal distribution with mean 0 and variance σ_e^2. Under this setup, the coverage probability of the ith observation is

$$
\pi_{\delta_0 i} = \Pr(|Y_i - X_i| < \delta_0)
$$

$$
= \Phi \left[\frac{\delta_0 - \beta_0 - (\beta_1 - 1)x_i}{\sigma_e} \right] - \Phi \left[\frac{-\delta_0 - \beta_0 - (\beta_1 - 1)x_i}{\sigma_e} \right].
$$

We define the overall coverage probability as

$$
\pi_{\delta_0 | X} = \frac{1}{n} \sum_{i=1}^n \pi_{\delta_0 i}. \tag{2.53}
$$

Suppose that we have a random sample $\{(Y_i, X_i) \mid i = (1, \ldots, n)\}$ and that β_0, β_1, and σ_e^2 are estimated by b_0, b_1, and s_e^2. Then b_0 and b_1 are independent of s_e. An estimate of $\pi_{\delta_0 i}$ is

$$
p_{\delta_0 i} = \Phi \left[\frac{\delta_0 - b_0 - (b_1 - 1)x_i}{s_e} \right] - \Phi \left[\frac{-\delta_0 - b_0 - (b_1 - 1)x_i}{s_e} \right],
$$

and an estimate of $\pi_{\delta_0 | X}$ is

$$
p_{\delta_0 | X} = \frac{1}{n} \sum_{i=1}^n p_{\delta_0 i}.
$$

By the same method as shown in Section 2.9.3, it can be shown that

$$
E(p_{\delta_0 | X}) = \pi_{\delta_0 | X} + O \left(\frac{1}{n} \right),
$$

and that the asymptotic variance of $p_{\delta_0 | X}$ is

$$
\sigma_{p_{\delta_0} | X}^2 = \frac{1}{n} \left[\frac{C_0^2}{n^2} + \frac{(C_0 \overline{X} - C_1)^2}{n^2 s_x^2} + \frac{C_2^2}{2n^2} \right] + O \left(\frac{1}{n^2} \right),
$$

where C_0, C_1 and C_2 are defined in (2.19), (2.20), and (2.21).

2.10.4   Accuracy Estimate

Here we use the same $\Sigma_{|X}$ as in (2.51). The elements of $\boldsymbol{d}_{|X} = \left(d_{|X1}, d_{|X2}, d_{|X3}\right)'$ become

$$d_{|X1} = -\frac{(\mu_y - \overline{X})^2}{\sigma_y s_x (1 - \chi_a)^2},$$

$$d_{|X2} = -\frac{s_x \beta_1}{\sigma_y (1 - \chi_a)^2} + \frac{s_x^2 \beta_1}{\sigma_y^2 \chi_a (1 - \chi_a)^2},$$

and

$$d_{|X3} = -\frac{1}{2\sigma_y s_x (1 - \chi_a)} + \frac{1}{2\sigma_y^2 \chi_a (1 - \chi_a^2)}.$$

Therefore, the variance of $L_{|X}$ is

$$\sigma_{L_{|X}}^2 = \frac{v^2 \varpi \chi_a^2 (1 - \rho^2) + \frac{1}{2}(1 - \varpi \chi_a)^2 (1 - \rho^4)}{n(1 - \chi_a)^2}.$$

2.11   Other Estimations and Statistical Inference Approaches

Estimations of agreement indices presented in this chapter are based on moment estimations by replacing proposed indices with their respective sample counterparts. Statistical inferences based on these estimates are carried out by the routine delta method.

King and Chinchilli (2001a) proposed estimations and statistical inferences for the CCC based on the U-statistic outlined by Davis and Quade (1968). Barnhart and Williamson (2001) proposed estimations and statistical inferences for the CCC based on the GEE methodology outlined by Liang and Zeger (1986). Barnhart, Haber, and Song (2002) later extended the GEE methodology for multiple assays or raters. Both U-statistic and GEE methodologies have the advantage of addressing both estimation and statistical inference simultaneously based on established general formulas.

Carrasco and Jover (2005) proposed to use the maximum likelihood (ML) method with a mixed effect model. All of the above approaches were proposed for CCC only, and have been extended to be applicable when we have multiple assays or raters. However, none of the above has addressed cases in which the target values are fixed. The U-statistic and GEE methodology for the CCC are also valid when we have categorical data, to be discussed in Chapters 3 and 5.

2.11.1 U-Statistic for CCC

Suppose we have $(X_1, Y_1), (X_2, Y_2), \ldots, (X_n, Y_n)$ random samples from n samples or subjects. For $i, j = 1, 2, \ldots, n$, let

$$\varphi_{1ij} = (X_i - Y_i)^2 + (X_j - Y_j)^2 - (X_i + Y_i)^2 - (X_j + Y_j)^2,$$

$$\varphi_{2ij} = X_i^2 + X_j^2 + Y_i^2 + Y_j^2,$$

$$\varphi_{3ij} = (X_i - Y_j)^2 - (X_i + Y_j)^2 + (X_j - Y_i)^2 - (X_j + Y_i)^2,$$

$$U_1 = \frac{\sum_{ij} \varphi_{1ij}}{2n(n-1)},$$

$$U_2 = \frac{\sum_{ij} \varphi_{2ij}}{n(n-1)},$$

and

$$U_3 = \frac{\sum_{ij} \varphi_{3ij}}{2n(n-1)}.$$

King and Chinchilli (2001a) showed that the CCC estimate can be written in terms of functions of the U-statistics as

$$r_c = \frac{H}{G} = \frac{(n-1)(U_3 - U_1)}{U_1 + nU_2 + (n-1)U_3}. \tag{2.54}$$

They further showed that the Z-transformation of r_c by the delta method has asymptotic normal distribution with mean $\frac{1}{2}\tanh^{-1}(\rho_c)$, and variance

$$\sigma_Z^2 = \frac{\rho_c^2}{\left(1 - \rho_c^2\right)^2}\left[\frac{\sigma_H^2}{H^2} - \frac{2\sigma_{HG}}{HG} + \frac{\sigma_G^2}{G^2}\right], \tag{2.55}$$

where

$$\sigma_H^2 = (n-1)^2[V(U_3) + V(U_1) - 2\text{cov}(U_3, U_1)],$$

$$\sigma_G^2 = (n-1)^2 V(U_3) + V(U_1) + n^2 V(U_2) + 2(n-1)\text{cov}(U_3, U_1)$$

$$+ 2n(n-1)\text{cov}(U_3, U_2) + 2n\text{cov}(U_3, U_2),$$

and

$$\sigma_{HG} = -(n-1)(n-2)\text{cov}(U_3, U_1) + n(n-1)\text{cov}(U_3, U_2) + (n-1)^2 V(U_3)$$

$$- (n-1)V(U_1) - n(n-1)\text{cov}(U_2, U_1).$$

The variance–covariance matrix of $U = (U_1, U_2, U_3)'$, denoted by V, can be obtained as follows. Let

$$\boldsymbol{\varphi}_{*i} = (\varphi_{1i}, \varphi_{2i}, \varphi_{3i})',$$

where

$$\varphi_{1i} = \frac{\sum_j \varphi_{1ij}}{(n-1)}, \quad \varphi_{2i} = \frac{\sum_j \varphi_{2ij}}{(n-1)}, \quad \varphi_{3i} = \frac{\sum_j \varphi_{3ij}}{(n-1)}.$$

Then we have

$$V = \frac{4}{n^2} \sum_i (\varphi_{*i} - U)'(\varphi_{*i} - U).$$

2.11.2   GEE for CCC

Barnhart and Williamson (2001) first proposed to use GEE for statistical estimation and inference for CCC. They used three sets of GEE equations, one for estimating means accounting for covariates, one for estimating variances without accounting for covariates, and one for estimating the Z-transformed CCC. Variance–covariance matrices of the above estimates can be estimated, and the delta method can be applied to obtain the asymptotic normality of the Z-transformed CCC estimate.

Suppose we have $(Y_{11}, Y_{21}), (Y_{12}, Y_{22}), \dots, (Y_{1n}, Y_{2n})$ random samples from n samples or subjects. Let Y_i be the 2×1 vector that contains the two readings and let the $2 \times p$ matrix X_i be the corresponding p covariates for the ith sample, where the first column of X_i is a vector of all ones representing an intercept term. Let $Y_i = (Y_{1i}, Y_{2i})'$ and β be a 2×1 marginal parameter vector. The three GEE equations are shown below.

In the first set of equations, the marginal mean vector of Y_i is $E(Y_i) = \mu_i = X_i \beta$, and the parameter estimates of β are obtained by

$$\sum_1^n D_i' V_i^{-1}(Y_i - \mu_i(\beta)) = 0, \qquad (2.56)$$

where $D_i = \frac{d\mu_i}{d\beta}$ and V_i is the working covariance matrix for Y_i (Zeger and Liang 1986).

Let σ_1^2 and σ_2^2 be variances of Y_1 and Y_2. In the second set of equations, the variances of Y_1 and Y_2 without accounting for covariates are estimated by

$$\sum_1^n F_i' H_i^{-1} \left(Y_i^2 - \sigma_i^2(\sigma^2, \beta)\right) = 0, \qquad (2.57)$$

where $F_i = \frac{d\sigma_i}{d\sigma^2}$ and H_i is the working covariance matrix for Y_i^2, $\sigma_i^2 = \sigma^2 + \mu_i^2$, and $\sigma^2 = (\sigma_1^2, \sigma_2^2)'$. In solving these equations, the diagonal components of H_i are assumed normal even if Y_i is not normally distributed.

Let $\theta_i = E(Y_{1i} Y_{2i})'$ and let Z be the Z-transformed CCC. In the third equation, Z is estimated by

$$\sum_1^n C_i W_i^{-1}(Y_1 Y_2 - \theta_i(Z, \beta, \sigma^2)) = 0, \qquad (2.58)$$

where $C_i = \frac{d\theta_i}{dZ}$ and W_i is the variance of θ_i.

This GEE method and the U-statistic method yield the same CCC estimate as proposed by Lin (1989), but the variances of the CCC estimate are slightly different because these two methods do not assume normality, while the method by Lin assumes normality in the computation of the variance of the CCC estimate.

2.11.3 Mixed Effect Model for CCC

Carrasco and Jover (2005) proposed to use the maximum likelihood (ML) or restricted ML (RML) method through a mixed effect model. Robieson (1999) and Carrasco and Jover (2005) showed that the CCC is a special form of ICC under the mixed effect model of random subject effect with the variance σ_α^2, residual effect with the variance σ_e^2, and fixed assay or rater effect with the mean square σ_β^2, when σ_β^2 is included in the denominator. Specifically, the CCC can be expressed as

$$\rho_c = \frac{\sigma_\alpha^2}{\sigma_\alpha^2 + \sigma_e^2 + \sigma_\beta^2}. \qquad (2.59)$$

In Section 3.1.3, we will revisit this coefficient in detail. Carrasco and Jover (2005) proceeded to use the delta method for statistical inference after estimating the variance–covariance matrix of the variance components through ML or RML.

This method does not yield the same CCC estimate as proposed by Lin (1989), and the variance of the CCC estimate can be slightly different, because this method assumes normality through MLE or RMLE.

2.11.4 Other Methods for TDI and CP

Compared to CCC, there have been fewer contributions related to TDI and CP. Some of the methods related to TDI and CP in the literature are pointed out in the last paragraph of Section 2.13. Most of those articles use more complicated iterative methods to fine-tune TDI and CP as well as their confidence intervals.

2.12 Discussion

2.12.1 Absolute Indices

Three agreement statistics, MSD, TDI, and CP, are unscaled indices, which do not depend on the between-subject variation. TDI and CP attempt to capture a large proportion (CP) of observations that are within a certain deviation (TDI) from their target values. We can compute CP for a given TDI, denoted by CP_{δ_0}, or compute TDI for a given CP, denoted by TDI_{π_0}. When the error structure is proportional, we apply a log transformation to the data, and the resulting TDI is then antilog transformed. When we subtract 1.00 from this antilog-transformed value and multiply by 100, it becomes a percent change ($TDI\%_{\pi_0}$) rather than an absolute difference (Lin 2000, 2003, Lin, Hedayat, Sinha, and Yang 2002). This means that $100\pi_0\%$ of observations are within $TDI\%_{\pi_0}$ of the target values. TDI and CP offer the most intuitively clear interpretation and have better power for statistical inference, yet they do not have precision and accuracy components. Also, Lin, Hedayat, Sinha, and Yang (2002) and Lin, Hedayat, and Wu (2007) used approximations and assumed normality to perform estimations and statistical inferences for TDI and CP. When the data are not normally or log-normally distributed, a transformation to bring the data closer to normality might be necessary.

TDI and CP are mirrored statistics. The former requires a given coverage probability to compute the absolute difference or percent change. The latter requires a given absolute difference or percent change to compute the coverage probability. The former has the advantage for the following reason:

- TDI can discriminate among assays with much better agreement than CP, because in these cases, CP values are near one.
- When there is no hard allowance available, one can still compute TDI at CP = 0.8 or 0.9, but one cannot compute CP without a reasonably given TDI value.

2.12.2 Relative Indices Scaled to the Data Range

Due to its equivalence to kappa and weighted kappa as well as its close tie to ICC, the CCC is perhaps the most popular index among statisticians for assessing agreement. The CCC and precision and accuracy coefficients are ICC-like (Lin, Hedayat, and Wu 2007), and are scaled (relative) to the total variation, especially the between-sample variation. This property is appealing when we wish to assess agreement over the entire reasonable value range from normal to abnormal. Comparisons among any of these three agreement statistics are valid only with similar study ranges, which is proportional to the between-sample variation. When the study range is fixed and meaningful, the CCC and precision and accuracy coefficients offer meaningful geometric interpretations. It is important to report the study range when reporting these statistics. Good agreement over a small range

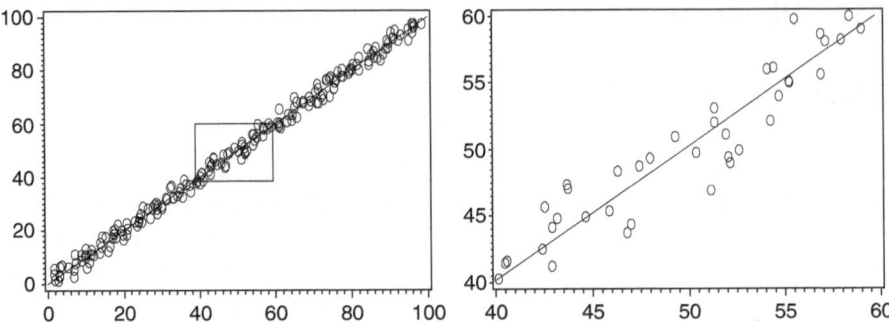

Fig. 2.13 Agreement over larger and shorter analytical ranges. (**a**) Larger analytical range. (**b**) Shorter analytical range

of measurements cannot be extrapolated to conclude good agreement over a larger range of measurements.

In other words, we should not conduct a method-comparison experiment over a range similar to the range of intrasample random fluctuation (Lin and Chinchilli 1997). As an illustrative example, Fig. 2.13a shows an artificial result of a good agreement over a desirable analytical range. If we study a subset of the data, limited to a much shorter analytical range (the box portion of Fig. 2.13a), which is magnified in Fig. 2.13b, any correlation coefficient would be much smaller.

2.12.3 Variances of Index Estimates Under Random and Fixed Target Values

Apart from CP, all of the agreement coefficients defined in this chapter have the same coefficient estimates regardless whether the target values are assumed random or fixed. The CP estimates under random and fixed target values are asymptotically the same. The variance of each of these coefficient estimates under the random target assumption is always larger than the variance under the fixed target assumption. These are evident by comparing (2.3) and (2.4) for the log MSD (TDI), (2.13) and (2.14) for the logit CP, (2.29) and (2.30) for the Z-transformed CCC, (2.31) and (2.32) for the logit accuracy coefficient, and the formulas in the text after (2.32) for the Z-transformed precision coefficient. Therefore, the confidence limits of an agreement coefficient would be closer to its coefficient estimate under the fixed target assumption than under the random target assumption.

2.12.4 Repeated Measures CCC

There are two types of repeated-measures data for agreement assessment that we often encounter. For one type, the between-sample variation forms the data range.

For the other, the repeated measures form the data range. An example of the first case is to have each subject's blood pressures measured over time, which is what is usually encountered in practice. Many tools are available for this type of repeated measure, which can be found in the Section 2.13. An example of the second case is to have each sample taken from a homogeneous population and to perform serial dilutions that form the data range. Such serial dilutions are uniform across all homogeneous samples. In this case, we could compute agreement coefficient estimates for each sample, and treat these estimates as random samples from a population. We then compute means and confidence limits based on the respective transformations of the agreement statistics. Antitransformation of these limits would be their respective confident limits. Such an approach is valid if we don't have missing data. We may also follow the more detailed approach proposed by Chinchilli, Martel, Kumanyika, and Lloyd (1996).

2.12.5   Data Transformations

The CCC and accuracy and precision coefficient estimates are in general quite robust against moderate deviation from normality. If not, there are tools based on robust estimates by King and Chinchilli (2001b) and based on a nonparametric approach by Guo and Manatunga (2007).

The TDI and CP estimates are heavily dependent on the normality or log-normality assumption. When there is evidence that data are not normally distributed for the constant random error case, and not log-normally distributed for the proportional random error case, data transformation might be necessary for the robustness of TDI and CP estimates. In this case, see Lin and Vonesh (1989) for the transformation approach that minimizes the MSD between the ordered observed and theoretical quantiles.

2.12.6   Missing Data

In this book we deal only with no-missing-data cases. In this chapter and Chapter 3, we discuss the case of paired assays or raters, and we often do not encounter a large amount of missing data in practice. Therefore, deleting cases with missing data is a reasonable approach. Missing data situation can sometimes be an issue in practice as pointed out in Chapters 5 and 6 when we have multiple raters with replicates. Research in the social and psychological sciences and in clinical trials may often encounter missing data. Approaches that can handle missing data should be an interesting area of research.

2.12.7 Account for Covariants

Covariate adjustment has been a controversial topic in assessing agreement. Proponents argue that without such adjustment, CCC and thus accuracy and precision coefficients are artificially inflated. Opponents argue that such adjustment artificially decreases the CCC because the data range is being reduced, as seen in an earlier paragraph about the effect of the data range. Opponents further argue that assay or rater agreement that could depend on covariants cannot be judged as having good agreement. In addition, often covariants are selected to cover a desirable data range, as in the use a variety of species in Example 2.8.5. In this case, the covariate adjustment related to species can be misleading. We believe that there are cases in which covariate adjustment is meaningful. However, we recommend that the practice of covariate adjustment be used with caution.

The GEE methodology proposed by Barnhart and Williamson (2001) allows for covariate adjustment, but only for means, not for variances and covariances. To adjust for variances and covariances, a simple and reasonable way is to perform the linear regression for each assay or rater, then use the intercept or overall mean estimate plus the residual of each assay or rater as the adjusted dependent variable. The adjusted dependent variable of each assay or rater would still contain the subject-to-subject variation without the effect of covariates. We can then perform the estimations and statistical inferences of the agreement indices based on these adjusted dependent variables. Such an approach is appropriate when covariates are subject-specific, such as age and gender. This approach is not applicable when we have fixed target values. However, we have rarely encountered covariates for which the target values are fixed.

2.13   A Brief Tour of Related Publications

Chapter 2 is based largely on the materials in Lin, Hedayat, Sinha, and Yang (2002). For an earlier introduction of CCC and precision and accuracy coefficients, see Lin (1989), and for TDI, see Lin (2000).

The method of Bland and Altman (1986) for assessing agreement uses a meaningful graphical approach and computes the confidence limits from the paired differences. Because of its simplicity, this approach has been quite popular among medical researchers. This approach lacks a specific index to summarize the degree of agreement, and thus statistical inferences about the estimate cannot be performed. Bland and Altman later (1999) improved on their approach with statistical inference. Their approaches are similar to our TDI. The major difference between our TDI and their approaches is that we capture a majority of observations from their respective individual target values, while their approaches capture the paired differences from the mean of paired differences.

Chinchilli, Martel, Kumanyika, and Lloyd (1996) addressed repeated-measures CCC, which is the weighted average of CCCs across subjects. Vonesh, Chinchilli, and Pu (1996) and Vonesh and Chinchilli (1997) modified CCC for goodness-of-fit. King and Chinchilli (2001a) used a U-statistics framework for CCC, which includes the generalization of CCC for multiple assays or raters, and the approach is applicable to categorical data. Barnhart and Williamson (2001) used GEE to estimate CCC, which also can be extended to include the generalization of CCC for multiple assays or raters (Barnhart, Haber, and Song 2002). In our opinion, these GEE methodologies are also applicable for categorical data, although the authors did not make such a claim.

King and Chinchilli (2001b) proposed a robust estimation of CCC through an absolute loss function or M-estimation by U-statistics. Li and Chow (2005) used weighted CCC by kernel density for repeated-measures image data. Quiroz (2005) proposed to assess agreement using the CCC in a repeated-measurement design. Liu, Du, Teresi, and Hasin (2005) proposed CCC for survival data. Barnhart, Song, and Lyles (2005) proposed assay validation for left-censored data. Carrasco and Jover (2005) proposed to use the maximum likelihood (ML) or restricted ML (RML) method through a mixed-effect model, which is applicable to multiple assays or raters. Robieson (1999) and Carrasco and Jover (2005) showed that the CCC is a special form of ICC under the mixed-effect model of random subject effect and residual effect and fixed assay or rater effect when the fixed assay or rater effect is included in the denominator of the ICC. King, Chinchilli, and Carrasco (2007) proposed another approach of repeated-measures CCC. They used the population estimates, rather than subject-specific estimates proposed earlier by Chinchilli, Martel, Kumanyika, and Lloyd (1996), to construct a repeated-measures CCC. King, Chinchilli, Wang, and Carrasco (2007) presented a class of repeated-measures CCC. Guo and Manatunga (2007) proposed using nonparametric estimation of the CCC under univariate censoring. Carrasco, Luis, King, and Chinchilli (2007) compared concordance correlation coefficient estimating approaches with skewed data. Williamson, Crawford, and Lin (2007) presented a permutation testing for comparing dependent CCCs. Quiroz and Burdick (2009) again proposed an assessment of individual agreement with repeated measurements based on generalized confidence intervals. Carrasco, King, and Chinchilli (2009) again proposed repeated-measures CCC estimated by variance components. Hiriote and Chinchilli (2010) proposed matrix-based concordance correlation coefficient for repeated measures. Helenowski, Vonesh, Demirtas, Rademaker, Ananthanarayanan, Gann, and Jovanovic (2011) extended CCC by allowing for different spatial variance–covariance structures of the data. They proposed a general concordance correlation matrix representing pairwise CCCs along with an overall CCC.

There have been relatively fewer publications for TDI and CP, which are summarized below. Wang and Gene Hwang (2001) proposed a nearly unbiased test (NUT) based on CP for the application of individual bioequivalence. Choudhary and Nagaraja (2007) proposed an exact test and modified NUT for CP and TDI for data with a small sample size (<30) that need to be solved numerically through iterations, and a bootstrap estimation for data with a moderate sample size. Hedayat, Lou, and

Sinha (2009) introduced CP involving multiple assays or raters. Escaramis, Ascaso, and Carrasco (2010) simplified the approach by Choudhary and Nagaraja (2007) for TDI using a tolerance limit approach through iterations. Choudhary (2008) proposed a tolerance interval approach for assessment of agreement in method comparison studies with repeated measurements. Choudhary (2008) proposed the tolerance approach with left censored data.

Chapter 3
Categorical Data

In Chapter 2, we discussed agreement statistics for continuous data in terms of absolute and relative indices. For categorical data, the agreement statistics presented here are quite different for random and fixed target values due to traditional practice over a long period of time. Therefore, the organization of this chapter is different from that of Chapter 2. In Section 3.1, we will discuss cases in which the target values are random. In Section 3.2, we will discuss cases in which the target values are fixed, which are rather simple and straightforward.

Agreement indices for categorical data have been conventionally regarded as measuring agreement among raters. Therefore, in this chapter, we refer to agreement among raters, observers, instruments, assays, methods, etc., as rater agreement, rather than as assay agreement as presented in Chapter 2.

3.1 Basic Approach When Target Values Are Random

There is a good collection of books and references describing agreement assessment for categorical data, including association and marginal agreement. A list of such references is given in Section 3.4, at the end of this chapter. We do not attempt to describe the details of all of available tools. Instead, we evaluate popular agreement indices for categorical data that are closely related to those described in Chapter 2. We will discuss the equivalence of ICC, CCC, and weighted kappa, and set the stage for unified approaches to be presented in Chapters 5 and 6 for continuous and categorical data when the target values are random.

3.1.1 Data Structure

Clinical measurements can be on a continuous or categorical scale. An example of the former is a patient's blood pressure. An example of the latter is the assignment

Table 3.1 Agreement
probability table

| | Category | \multicolumn{4}{c|}{Y} | |
		1	2	...	t	Total
	1	π_{11}	π_{12}	...	π_{1t}	$\pi_{1\cdot}$
	2	π_{21}	π_{22}	...	π_{2t}	$\pi_{2\cdot}$
X	⋮	...	...	...	...	⋮
	t	π_{t1}	π_{t2}	...	π_{tt}	$\pi_{t\cdot}$
	Total	$\pi_{\cdot 1}$	$\pi_{\cdot 2}$	...	$\pi_{\cdot t}$	1

by two or more raters of a patient's condition according to an ordinal scale of fair, mild, serious, critical, or life-threatening, or the assignment of a patient's condition to a binary scale of normal or abnormal. Here, the raters should use the same metrics in judging. Evaluation of agreement for data with a nominal scale without any order will be discussed in Section 3.3.

We begin with the most basic scenario for paired ordinal observations (Y and X) with a bivariate multinomial distribution when both Y and X take on values of 1 to t, or 0 to $t - 1$. Table 3.1 presents the agreement data for all possible probability outcomes, where π_{ij} represents the probability of $X = i$ and $Y = j, i, j = 1, 2, \ldots, t$.

3.1.2 Absolute Indices

Absolute agreement indices for categorical data have rarely been presented in the published literature. However, these can be very useful when we try to verify agreement within a given experiment. For example, when replicates are taken per rater per subject, we might want to compare whether one rater is better than another in terms of within-rater precision. For another example, in the field of individual bioequivalence, we might be interested in evaluating interdrug agreement relative to intradrug agreement. These scenarios will be presented in Chapter 6, when we concentrate on comparisons of absolute indices rather than of relative indices.

We have no knowledge whether MSD has been used for categorical data in the literature. We defined MSD in Chapter 2 as $\varepsilon^2 = E(Y - X)^2$. Based on agreement probabilities presented in Table 3.1, MSD becomes

$$\varepsilon^2 = \sum_i^t \sum_j^t (i - j)^2 \pi_{ij}, \quad i, j = 1, 2, \ldots, t. \tag{3.1}$$

When the outcome takes on binary scores, or $t = 2$, the MSD becomes $\pi_{12} + \pi_{21}$, or $1 - (\pi_{11} + \pi_{22})$. The MSD shown in (3.1) is the absolute or unscaled measure of disagreement. The absolute measure of agreement, Π_0, can be defined as a weighted average of all probabilities,

$$\Pi_0 = \sum_i^t \sum_j^t w_{ij} \pi_{ij}; \tag{3.2}$$

a popular weight function, the squared weight function, is

$$w_{ij} = 1 - \frac{(i-j)^2}{(t-1)^2}. \tag{3.3}$$

Equation (3.2) with the weight function (3.3) is actually a linear function of MSD through the relationship

$$\Pi_0 = 1 - \frac{\varepsilon^2}{(t-1)^2}. \tag{3.4}$$

The weight function in (3.3) assigns the heaviest weight of 1 when $i = j$ for the main diagonal probabilities, and lower weights depending on the squared distance from the main diagonal probabilities. We will discuss the weight function in greater detail in Section 3.1.3. If we let

$$w_{ij} = \begin{cases} 1 & \text{when } i = j, \\ 0 & \text{otherwise,} \end{cases}$$

then we have $\Pi_0 = \sum_i \pi_{ii}$, which had been commonly used as an agreement index prior to the introduction of kappa by Cohen (1960). When the outcome takes on binary scores, or $t = 2$, Π_0 becomes $\pi_{11} + \pi_{22}$.

3.1.3 Relative Indices: Kappa and Weighted Kappa

The controversy associated with the absolute indices such as Π_0 is that a certain amount of agreement is to be expected by chance alone. Even when both raters have totally different metrics in judging the subjects, we would expect them to agree with a certain probability Π_c by chance alone. It is natural to eliminate such chance agreements. This is the idea behind Cohen's kappa (1960).

To better illustrate the arguments made above, let us look at some numerical examples when $t = 2$ and $n = 100$:

1. Perfect agreement: Raters 1 and 2 agree on every subject in the following table. There is no controversy expected to conclude the perfect agreement between the two raters.

		Rater 2		
		Yes	No	Total
	Yes	50	0	50
Rater 1	No	0	50	50
	Total	50	50	100

2. No agreement: Raters 1 and 2 agree on $25 + 25 = 50$ of the 100 subjects they examined in the following table. Although these two raters agree 50% of the

time, in fact, there is no association between the raters in this example because all their agreement can be attributed to chance.

		Rater 2		
		Yes	No	Total
	Yes	25	25	50
Rater 1	No	25	25	50
	Total	50	50	100

Also note that in each of the above examples, the marginal distributions are the same for the two raters. Just as for continuous data, the fact of identical marginal distributions between the two raters does not necessarily imply agreement between them.

Cohen (1960) proposed a measure called *kappa* as a chance-corrected agreement index. This index depends strictly on the diagonal probabilities corrected by the chance probabilities derived from the marginal probabilities.

Cohen (1968) later improved on the kappa coefficient by proposing *weighted kappa* for data measured in ordinal scales. We will discuss weighted kappa first, because kappa is a special case of weighted kappa. Weighted kappa is designed to recognize that some disagreements between the two raters should be considered more serious than others. For example, disagreement between "mild" and "life-threatening" for a patient's condition is more serious than disagreement between "critical" and "life-threatening." Therefore, it would be prudent to assign weights to reflect the seriousness of disagreements among the rated conditions. In general, for assessing agreement (disagreement), we would expect the weight to be greater (smaller) for cells closer to (farther from) the main diagonal.

The weighted measure of agreement by chance becomes

$$\Pi_c = \sum_i^t \sum_j^t w_{ij} \, \pi_{i.} \pi_{.j},\tag{3.5}$$

where $\pi_{i.}(\pi_{.j})$ represents the marginal probability by rater X (rater Y). When the outcome takes on binary scores, or $t = 2$, then Π_c reduces to $\pi_{1.}\pi_{.1} + \pi_{2.}\pi_{.2}$.

Finally, weighted kappa is defined as

$$\kappa_w = \frac{\Pi_0 - \Pi_c}{1 - \Pi_c}.\tag{3.6}$$

Cohen requires these weights to satisfy the following conditions:

1. $w_{ij} = 1$ when $i = j$,
2. $0 < w_{ij} < 1$ when $i \neq j$,
3. $w_{ij} = w_{ji}$.

Cicchetti and Allison (1971) suggested the following set of weights:

$$w_{ij} = 1 - \frac{|i - j|}{t - 1}.\tag{3.7}$$

Fleiss and Cohen (1973) suggested the squared weight function as shown in (3.3). Weighted kappa based on the weight function in (3.7) is always less than or equal to that based on the weight function in (3.3).

When

$$w_{ij} = \begin{cases} 1 & \text{for } i = j, \\ 0 & \text{for } i \neq j, \end{cases}$$

weighted kappa becomes the kappa coefficient originally proposed by Cohen (1960). This kappa coefficient has been used for data with a binary or nominal scale. Weighted kappa, regardless of which weight function is used, degenerates into kappa in the binary case. These relative agreement indices are invariant under any linear transformation on the paired data.

3.1.4 Sample Counterparts

Suppose we have two raters each of whom evaluates and assigns n subjects independently to one of t categories. Let p_{ij} represent the proportion of subjects assigned to category i by rater X and category j by rater Y, where $i, j = 1, 2, \ldots, t$. Let $p_{i\cdot}$ ($p_{\cdot j}$) represent the marginal proportion that a subject is assigned to category i (category j) by rater X (rater Y), where $i, j = 1, 2, \ldots, t$. Then the expected values of the above proportions are $E(p_{ij}) = \pi_{ij}$, $E(p_{i\cdot}) = \pi_{i\cdot}$, and $E(p_{\cdot j}) = \pi_{\cdot j}$. These proportions are also the maximum likelihood estimators (MLEs) or sample moment counterparts of the respective probabilities.

3.1.5 Statistical Inference on Weighted Kappa

Fleiss, Cohen, and Everitt (1969) proposed to estimate weighted kappa using the sample counterpart as

$$\hat{\kappa}_w = \frac{P_0 - P_c}{1 - P_c}, \tag{3.8}$$

where

$$P_0 = \sum_i^t \sum_j^t w_{ij}\, p_{ij}, \quad P_c = \sum_i^t \sum_j^t w_{ij}\, p_{i\cdot} p_{\cdot j}.$$

The weighted kappa estimate $\hat{\kappa}_w$ has an asymptotic normal distribution with mean κ_w and variance

$$\sigma_{\hat{\kappa}_w}^2 = \frac{\sum_i \sum_j \pi_{ij}\left[w_{ij} - (\bar{w}_{i\cdot} + \bar{w}_{\cdot j})(1 - \kappa_w)\right]^2 - \left[\kappa_w - \Pi_c(1 - \kappa_w)\right]^2}{n(1 - \Pi_c)^2}, \tag{3.9}$$

where $\bar{w}_{i\cdot} = \sum_j \pi_{\cdot j} w_{ij}$ and $\bar{w}_{\cdot j} = \sum_i \pi_{i\cdot} w_{ij}$.

Statistical inference can be performed using sample counterparts for (3.9), denoted by $s_{\hat{k}_w}$. We can compute a 95% one-tailed lower confidence limit as

$$\hat{k}_w - 1.645 s_{\hat{k}_w}.$$

These Cohen's kappa coefficients have been widely used and can be obtained from the output of SAS procedure FREQ if option "agree" is used in the TABLES statement.

3.1.6   Equivalence of Weighted Kappa and CCC

Using the weight function defined in (3.3), we can verify that the weighted kappa is actually equal to the CCC presented in Chapter 2. Recall that CCC was defined in (2.24) as

$$\rho_c = 1 - \frac{\varepsilon^2}{\varepsilon^2|_{\rho=0}}.$$

When $\rho = 0$, MSD becomes

$$\varepsilon^2|_{\rho=0} = \sum_i^t \sum_j^t (i - j)^2 \pi_{i.} \pi_{.j}, \quad i, j = 1, 2, \ldots, t. \tag{3.10}$$

Therefore, (3.5) becomes

$$\Pi_c = \sum_i^t \sum_j^t w_{ij} \pi_{i.} \pi_{.j} = 1 - \frac{\varepsilon^2|_{\rho=0}}{(t-1)^2}.$$

Together with (3.4), the CCC becomes

$$\rho_c = 1 - \frac{1 - \Pi_0}{1 - \Pi_c} = \frac{\Pi_0 - \Pi_c}{1 - \Pi_c}, \tag{3.11}$$

which is exactly the same as the weighted kappa in (3.6).

   The equivalence of weighted kappa using the squared distance weight function and CCC can also be found in Robieson (1999), King and Chinchilli (2001a, 2001b), and Barnhart, Haber, and Song (2002). King and Chinchilli (2001a, 2001b) proved that the weighted kappa estimate using the weight function (3.7) actually is exactly the same as the CCC estimate with the absolute difference function by U-statistic.

   The variance presented in (3.9) is not the same as the variance of the CCC estimate presented in (2.29), because that the CCC presented in Chapter 2 uses the normality assumption in deriving the variance of the CCC estimate. In Chapter 5 we will prove that the equivalence can be established using general estimating equations (GEE) without assuming normality.

3.1.7   Weighted Kappa as Product of Precision and Accuracy Coefficients

For the agreement statistics presented in Sections 3.1.2 and 3.1.3, two raters need to classify subjects or samples based on the same metrics. When the metrics are not the same, or when the shifts in marginal distributions are negligible, then we are dealing with association instead of agreement. Several statistical tools are available for measuring and evaluating association, which are listed in Section 3.4. Popular association measurements include the Pearson correlation coefficient, χ^2 statistic for contingency tables, some forms of intraclass correlation coefficients (ICC), and the log-linear modeling, among several others. If we use the Pearson correlation coefficient for measuring association, we can regard it as a precision measurement as in the CCC. Accuracy can be characterized by examining the differences in the marginal distributions. Such statistical tools are also widely available in the textbooks listed in Section 3.4. In this book we concentrate on the accuracy and precision topics that mirror those of CCC in Section 2.3.

The equivalence of CCC and weighted kappa can shed new light on the way we view the weighted kappa. We can break κ_w into precision (Pearson correlation coefficient, ρ) and accuracy (χ_a) coefficients in the same way we did for CCC. That is,

$$\kappa_w = \rho \chi_a, \tag{3.12}$$

where

$$\rho = \frac{\sigma_{yx}}{\sigma_y \sigma_x} \tag{3.13}$$

and

$$\chi_a = \frac{2\sigma_y \sigma_x}{\sigma_y^2 + \sigma_x^2 + (\mu_y - \mu_x)^2}. \tag{3.14}$$

Here,

$$\mu_x = \sum_i^t i \pi_{i\cdot},$$

$$\mu_y = \sum_j^t j \pi_{\cdot j},$$

$$\sigma_x^2 = \sum_i^t i^2 \pi_{i\cdot} - \left(\sum_i^t i \pi_{i\cdot} \right)^2,$$

$$\sigma_y^2 = \sum_j^t j^2 \pi_{\cdot j} - \left(\sum_j^t j \pi_{\cdot j} \right)^2,$$

and

$$\sigma_{yx} = \sum_i^t \sum_j^t ij\pi_{ij} - \left(\sum_i^t i\pi_{i\cdot}\right)\left(\sum_j^t j\pi_{\cdot j}\right).$$

We can see that ρ depends on cell probabilities, while χ_a strictly depends on marginal probabilities. In the most basic scenario of $i, j = 1, 2$, the precision coefficient becomes

$$\rho = \frac{\pi_{11}\pi_{22} - \pi_{12}\pi_{21}}{\sqrt{\pi_{1\cdot}\pi_{2\cdot}\pi_{\cdot 1}\pi_{\cdot 2}}}, \tag{3.15}$$

and the accuracy coefficient is defined as

$$\chi_a = \frac{2\sqrt{\pi_{1\cdot}\pi_{2\cdot}\pi_{\cdot 1}\pi_{\cdot 2}}}{\pi_{1\cdot}\pi_{2\cdot} + \pi_{\cdot 1}\pi_{\cdot 2} + (\pi_{2\cdot} - \pi_{\cdot 2})^2}. \tag{3.16}$$

Obviously, $0 < \chi_a \leqslant 1$, and $\chi_a = 1$ when the marginal probabilities are identical. Since $\kappa_w = \rho\chi_a$, the weighted kappa is uniquely determined by the marginal probabilities and the Pearson correlation coefficient, even though the cell probabilities could vary. Inferences on these accuracy and precision coefficients for categorical data will be discussed in Chapter 5.

3.1.8 Intraclass Correlation Coefficient and Its Association with Weighted Kappa and CCC

CCC is a special form of ICC (Robieson 1999, Carrasco and Jover 2003) when the mean square of the difference among raters (fixed rater effect) is included in the denominator. Therefore, CCC and weighted kappa can be estimated by variance components with related statistical inferences through the delta method.

For simplicity, we first demonstrate the equivalence of ICC and CCC based on the basic mixed effect model where each of n subjects is evaluated by two raters. The value of Y_{ij} is the category that rater j had assigned to subject i. We will discuss a more general form of ICC in Chapter 5, where we have $k \geqslant 2$ raters, each evaluating n subjects with $m \geqslant 1$ readings.

The mixed effect model considered here is

$$Y_{ij} = \mu + \alpha_i + \beta_j + e_{ij}, \quad j = 1, 2, \ i = 1, 2, \dots, n, \tag{3.17}$$

where the overall mean is μ. The fixed rater effect is β_j and sums to zero. The random subject effect is α_i with the variance σ_α^2. The random error is e_{ij} with variance σ_e^2, and e_{ij} is assumed not to correlate with α_i.

Traditionally, ICC in its original form is defined as the between-subject variance σ_α^2 divided by the total variance of $\sigma_\alpha^2 + \sigma_e^2$ in the above model, disregarding the rater effect β_j. This traditional ICC can be interpreted in terms of rater consistency or reliability rather than rater agreement (Shrout and Fleiss 1979). The expected value of an individual observation in model (3.17), regardless of data being continuous, ordinal, or binary, is

$$E(Y_{ij}) = \mu_j = \mu + \beta_j, \quad j = 1, 2.$$

The covariances among individual observations are

$$\text{cov}(Y_{ij}, Y_{i'j'}) = \begin{cases} \sigma_\alpha^2 + \sigma_e^2, & \text{if } i = i', j = j', \\ \sigma_\alpha^2, & \text{if } i = i', j \neq j', \\ 0, & \text{if } i \neq i'. \end{cases} \quad (3.18)$$

For $j = 2$, we have $\sigma_\alpha^2 = \sigma_{12}$, or the covariance of the two raters, and

$$\sigma_\alpha^2 + \sigma_e^2 = \frac{\sigma_1^2 + \sigma_2^2}{2}, \quad (3.19)$$

where σ_j^2 is the variance of the effect associated with rater j, $j = 1, 2$.

Let σ_β^2 be the rater mean squares, which is

$$\sigma_\beta^2 = \frac{(\mu_1 - \mu_2)^2}{2}.$$

The CCC or weighted kappa becomes

$$\rho_c = \kappa_w = \frac{2\sigma_{12}}{\sigma_1^2 + \sigma_2^2 + (\mu_1 - \mu_2)^2} = \frac{\sigma_\alpha^2}{\sigma_\alpha^2 + \sigma_e^2 + \sigma_\beta^2}.$$

Model (3.17) assumes equal variances between the two raters. It is interesting to note that the CCC remains the same whether the variances are assumed equal or not. It is evident from (3.19) that $\sigma_\alpha^2 + \sigma_e^2$ is actually the average of the two rater variances. However, the composition of accuracy and precision coefficients are slightly altered under this model. We will discuss the ICC and CCC in greater detail in Chapter 5.

3.1.9 Rater Comparison Example

This example is taken from von Eye and Schuster (2000). Two psychiatrists evaluated 129 patients who had previously been diagnosed as clinically depressed. The rating categories were 0 for not depressed, 1 for mildly depressed, and 2 for clinically depressed. Table 3.2 presents the rating results that the two psychiatrists provided.

Table 3.3 presents the related agreement statistics and their confidence limits for the data reported in Table 3.2. There is a moderate level of agreement (weighted kappa = 0.4204 with a 95% lower confidence limit of 0.2737 by the squared weight function) between the two psychiatrists with good accuracy and moderate precision.

Table 3.2 Severity of
depression evaluated by two
psychiatrists

	Category	Rater Y			Total
		0	1	2	
	0	11	2	19	32
	1	1	3	3	7
Rater X	2	0	8	82	90
	Total	12	13	104	129

Table 3.3 Agreement
statistics on severity of
depression evaluated by two
psychiatrists

Statistics	Estimate	95% LCL[a]
Kappa (with 0 or 1 weights)	0.3745	0.2448
Weighted kappa[b]	0.4018	0.2653
Weighted kappa[c]	0.4204	0.2737
Pearson correlation	0.4694	0.3369
Spearman correlation	0.4202	0.2742

[a]LCL: lower confidence limit
[b]weighted by Ciccetti and Allison function, (3.7)
[c]weighted by Fleiss and Cohen function, (3.3)

3.2   Basic Approaches When Target Values Are Fixed: Absolute Indices

When the target values are fixed, investigators rarely examine results using a relative index. We consider the example of a diagnostic test. We select n_0 negative samples and n_1 positive samples to be tested as negative (0) or positive (1) by an instrument or a rater. Here, n_0 and n_1 are known. Fleiss (1973) illustrated this type of data in Section 1.2 of his book. Let us consider the 2×2 table presented in Table 3.4, from n_0 and n_1 negative and positive samples for the assessment of sensitivity and specificity in Section 3.2.1.

3.2.1   Sensitivity and Specificity

Sensitivity is the conditional probability of positive response given that the sample is positive, $P(Y = 1 \mid X = 1)$, which can be estimated by $p_{ss} = n_{11}/n_1$. The larger the value of $P(Y = 1 \mid X = 1)$, the more sensitive the test. Specificity is the conditional probability of negative response given that the sample is negative, $P(Y = 0 \mid X = 0)$, which can be estimated by $p_{sp} = n_{00}/n_0$. The larger the value of $P(Y = 0 \mid X = 0)$, the more specific the test. These parameters can be directly estimated from the 2×2 table presented in Table 3.4. Statistical inference related to these parameters can simply be addressed by the inference on related proportions. Variances of sensitivity and specificity can be estimated to be $p_{ss}(1 - p_{ss})/n_1$ and

Table 3.4 2×2 table under fixed sample sizes

		Observed Y		
	Category	0	1	
Actual X	0	n_{00}	n_{01}	n_0
	1	n_{10}	n_{11}	n_1

$p_{sp}(1 - p_{sp})/n_0$, respectively. When n_0 or n_1 is large (say ≥ 30), we can use normal approximation to compute a 95% confidence interval as

$$p_{ss} \pm 1.96 \sqrt{\frac{p_{ss}(1 - p_{ss})}{n_1}} \tag{3.20}$$

and

$$p_{sp} \pm 1.96 \sqrt{\frac{p_{sp}(1 - p_{sp})}{n_0}}, \tag{3.21}$$

respectively. Here, we are often interested in only the one-tailed 95% lower confidence limit.

When n_0 or n_1 is not large enough, we can use the binomial (exact) confidence interval approach. We take the inverse of the binomial distribution with parameters n_0 or n_1, p_{ss} or p_{sp}, and with the probability 0.025 for the lower limit and 0.975 for the upper limit. For the one-tailed lower limit, we use the probability 0.05. We then divide these limits by either n_0 or n_1 for the confidence limits on sensitivity and specificity.

3.2.1.1 False Positive Rate and False Negative Rate

Epidemiologists distinguish two types of error rates. The false positive error rate, π_{F+}, is defined as the proportion of negative samples among those tested positive. The false negative error rate, π_{F-}, is defined as the proportion of positive samples among those tested negative. These error rates are typically defined through the application of Bayes's theorem by

$$\pi_{F+} = P(X = 0 \mid Y = 1) = \frac{P(Y = 1 \mid X = 0)P(X = 0)}{P(Y = 1)} \tag{3.22}$$

and

$$\pi_{F-} = P(X = 1 \mid Y = 0) = \frac{P(Y = 0 \mid X = 1)P(X = 1)}{P(Y = 0)}. \tag{3.23}$$

These error rates cannot be directly estimated by the 2×2 table presented in Table 3.4, because the sample sizes, n_0 and n_1, are fixed in the study. In order to estimate the above error rates, we will need to collect the information about the true positive (disease) rate, $P(X = 1)$. Here, $P(X = 0)$ is equal to $1 - P(X = 1)$.

Table 3.5 Skin cancer diagnosed by pathological evaluation and by the preresection evaluation of a dermatologist

		Pre-resection		
	Category	No	Yes	
Pathological	No	112	6	118
	Yes	10	63	73

Table 3.6 Agreement statistics between preresection and pathological evaluations

Statistics	Estimate	95% LCLb[a]	95% LCLn[b]
Sensitivity	63/73 = 0.863	0.795	0.797
Specificity	112/118 = 0.949	0.915	0.916

[a]LCLb: lower confidence limit by binomial distribution
[b]LCLn: lower confidence limit by normal approximation

For assay validation or diagnostic lab testing, there is little interest in estimating these error rates in a population. Only the sensitivity and specificity values are of interest here. However, in the diagnostic environment, it is common practice to regard 1-sensitivity as the false negative error rate and 1-specificity as the false positive error rate.

3.2.2 Diagnostic Test Example

This example is taken from Linn (2004). In a case-control study, pathological diagnoses of skin cancer ($n_1 = 73$) and benign tumors ($n_0 = 118$) (defining the "true disease status") were recorded. Patients were also evaluated by the preresection clinical diagnostic (the "test") by a dermatologist. Table 3.5 presents the rating results that the two raters provided.

Table 3.6 presents the related agreement statistics and their associated confidence limits. The dermatologist has a good specificity (0.949 with a 95% lower confidence limit of 0.915) and a moderate sensitivity (0.863 with a 95% lower confidence limit of 0.795). Note that in this case-control study, the sample sizes of case ($n_1 = 73$) and control ($n_0 = 118$) were fixed. Therefore, the prevalence of the disease (skin cancer) cannot be estimated from this study.

3.3 Discussion

We have examined the agreement statistics for ordinal and binary data when target values are fixed or random. When target values are fixed, sensitivity and specificity coefficients are commonly used under the fixed n_0 and n_1 samples. We rarely conduct a study from n pairs of observations with fixed target values without fixing the n_0 and n_1 samples.

When target values are random, kappa and weighted kappa are very popular indices for measuring agreement for binary and ordinal data. The weighted kappa with the squared weight function is identical to the CCC introduced in Chapter 2. In Chapter 5, we will show that using the GEE methodology, the confidence limits of CCC and weighted kappa or kappa are identical when the Z-transformation is not used. When data have a nominal scale such as "depression, personality disorder, schizophrenia, and others," use of the simple kappa with diagonal weights of one and zero otherwise has been common practice. Fleiss (1971) introduced category-specific kappa for nominal data, which is more in-depth with examination of the agreement–disagreement matrix among the nominal categories.

The U-statistic approach in Section 2.11.1 is another novel approach to estimating weighted kappa with statistical inference. In this approach, the weight function of Cicchetti and Allison (1971) given in (3.7) can be applied. This approach allows for extension to the multiple-raters case. The GEE methodology proposed by Barnhart and Williamson (2001) can also be applied to estimate weighted kappa with the squared weight function with statistical inference. Both the U-statistic and GEE methodologies are applicable when the target values are random.

CCC or weighted kappa is a special form of ICC when the mean square difference among the fixed rater effect is included in the denominator, which can be estimated by variance components with statistical inference through the delta method.

The data structure discussed in this chapter has focused on the most basic case, namely, two raters evaluating each of n subjects only once. In Chapters 5 and 6, we will present and discuss tools for cases of at least two raters evaluating each of n subjects, and each could have $m \geq 1$ replicates. Such tools can be used for both categorical and continuous data when the target values are random.

In mental health studies, numerous instruments have been developed for the diagnosis of psychiatric disorders such as major depression, and there is considerable interest in replacing one instrument by another instrument for reduction of cost, ease of administration, and other considerations.

However, since these instruments are based on different questionnaires with distinctive structures and point systems, they often have different scales. When the scales of the instruments are different, the existing agreement methodology is not applicable. For example, in a depression study, depression was measured by a clinician-administered ordinal scale of no depression, mild depression, and severe depression, and by a continuous scale of dimensional self-report. It remains unknown whether the less-time-consuming self-report dimensional scale can replace clinician-administered scale to determine the grade of the illness. This problem is equivalent to assessing the extent to which the continuous scale can be interpreted as the ordinal graded severity of depression.

Due to the different measurement scales, this question cannot be addressed in the classical framework of agreement. Alternatively, we may perform a jackknife (leaving one out) discriminant analysis to best classify the continuous scale against the ordinal scale of no depression, mild depression, and severe depression. We can then assess the weighted kappa of the classified scale based on the optimal cutoff values of the continuous scale against the more definitive clinician-administered ordinal scale.

As revealed in Shoukri (2004), in the evaluation of diagnostic tests it is well known that certain tests that appear to have high sensitivity and specificity may on the other hand have low predictive accuracy when the prevalence of the disease is low. When the prevalence rate is low, we would have an asymmetric cluster of the observations observed in the nondisease category and with sparse cell observations otherwise. This situation is similar to the case of continuous data when the data range is short and with a few outliers. This leads to a higher probability of agreement by chance and therefore a lower kappa value. In this case, in analogy to the use of TDI for continuous data, perhaps we should also present the absolute (unscaled) agreement index as shown in (3.2).

3.4   A Brief Tour of Related Publications

For the assessment of bias among raters with respect to marginal distributions, the literature includes McNemar's test (1947), Cochran's Q test (1950), Madansky's Q test (1963), and Friedman's χ^2 test (1937). In addition, Fleiss and Everitt (1971) proposed a method for comparing the marginal distributions of an agreement table, and Koch, Landis, Freeman, Freeman Jr, and Lehnen (1977) proposed marginal homogeneity tests within the multivariate categorical data. Darroch (1981) introduced Mantel–Haenszel tests for marginal symmetry. Landis, Sharp, Kuritz, and Koch (1998) proposed generalized Mantel–Haenszel tests for both nominal and ordinal data scales. Such bias assessment can also be captured in the accuracy coefficient, as shown in (3.14). These correspond to the accuracy component of weighted kappa.

For the assessment of association, the Pearson correlation coefficient and the usual χ^2 test of association have been available for many decades. Such association assessment can also be captured in the precision coefficient, as shown in (3.13). Birch (1964, 1965) proposed partial association under 2×2 and general cases. For the assessment of reliability, the ICC in its original form (Fisher 1925) is the ratio of between-sample variance and total (within plus between) variance under the model of equal marginal distributions. This original ICC was intended to measure precision only.

Several forms of ICC have evolved. In particular, Bartko (1966), Shrout and Fleiss (1979), Fleiss (1986), and Brennan (2001) have put forth various reliability assessments. Landis and Koch (1977b) introduced category-specific intraclass and interclass correlation coefficients within a multivariate variance-components model for categorical data that can accommodate unbalanced designs. These correspond to the precision component of weighted kappa. In Chapter 5, we will address various forms of ICCs, including the one that represents precision.

For the assessment of agreement, Cohen (1960) introduced the kappa coefficient to measure agreement between two raters on a nominal categorical data scale, followed by Cohen (1968) and Everitt (1968) each separately proposing a weighted kappa coefficient. However, Fleiss noted that the proposed variance estimators for both of these agreement measures were incorrect, and invited both Cohen and Everitt to collaborate in publishing correct variances, which appeared in Fleiss, Cohen, and Everitt (1969). Fleiss (1971) introduced category-specific kappa coefficients and generalized Cohen's kappa to situations involving multiple observers and multiple categories.

Combining these broad areas into a common estimation and hypothesis-testing framework, Landis and Koch (1977a) proposed the multivariate categorical data framework of Koch, Landis, Freeman, Freeman Jr, and Lehnen (1977) for the analysis of repeated measurement designs. The focus of this framework was to test first-order marginal homogeneity among multiple observers and to estimate multiple correlated kappa coefficients and their associated estimated covariances to facilitate ease of confidence interval construction. Cicchetti and Fleiss (1977) studied the null distributions of weighted kappa and the C ordinal statistic. Fleiss and Cuzick (1979) further proposed kappa for binary response when the number of observers differs for each subject. Bloch and Kraemer (1989) discussed 2×2 kappa coefficients as measures of agreement or association in greater details. Donner and Eliasziw (1992) proposed a goodness-of-fit approach to inference procedures and sample size estimation for the kappa statistic. Shoukri and Martin (1995) proposed MLE of the kappa coefficient from models of matched binary responses. Williamson, Lipsitz, and Manatunga (2000) proposed modeling kappa for measuring dependent categorical agreement data. Schuster (2001) used kappa as a parameter of a symmetry model for rater agreement. Broemeling (2009) proposed Bayesian methods for measure of agreement.

Another tool for assessing agreement is the log-linear model for rater agreement of von Eye and Schuster (2000). There is a wealth of books (Landis and Koch 1977c; Haberman 1974, 1978; Goodman 1978; Haberman 1979; Fleiss, Levin, Paik, and Wiley (1981); Aickin 1983; Freeman 1987; Cox and Snell 1989; Shoukri and Edge 1996; Christensen 1997; Agresti 1990; Von Eye and Mun 2005, to name a few) describing agreement assessment, including association, for categorical data. Furthermore, Haber, Gao, and Barnhart (2007) introduced a method for assessing observer agreement in studies involving replicated binary observations. Guo and Manatunga (2009) proposed a method of measuring agreement of multivariate discrete survival times using a modified weighted kappa coefficient. Yang and Chinchilli (2009, 2011) proposed fixed-effects modeling of Cohen's kappa for bivariate multinomial data.

Chapter 4
Sample Size and Power

In this chapter, in addition to the formal method of computing the sample size and power, we will discuss ways to compute sample size and power based on normal approximation through a simplified and conservative approach. As pointed out previously in assessing agreement, the traditional null and alternative hypotheses should be reversed. The conventional rejection region is now the region of declaring agreement, and is usually one-sided. Asymptotic power and sample size calculations will proceed by this principle.

4.1 The General Case

We will present the asymptotic power of accepting agreement and sample size calculation whereby we utilize either MSD or CCC in accepting or rejecting agreement. Inference based on approximated TDI or CP can be assessed through MSD. Let λ be a transformed agreement index. For continuous data, we use the Z-transformation for a CCC estimate, the logit transformation for a CP estimate, and the natural log transformation for an MSD estimate. When λ is MSD, we declare that two raters are in agreement under the alternative hypothesis $H_0 : \lambda < \lambda_0$, where λ_0 is a prespecified tolerable index value or the null value. Here, the null hypothesis is $H_0 : \lambda \geq \lambda_0$. When λ is CCC, we reverse the signs of the above hypotheses. We compute the probability of declaring agreement under the alternative value λ_1, which can be the ideal condition value or the available historical value. We refer to this probability as power. Lin, Hedayat, Sinha, and Yang (2002) presented the power and sample size computations that are described in the sequel.

Let η_0^2/n_c and η_1^2/n_c be variances of the respective estimates, where $n_c = n - c$, with $c = 2$ for MSD and CCC, and $c = 3$ for CP. For the one-tailed fixed type-I error α, the power for declaring agreement based on CCC and CP becomes

$$P = 1 - \Phi\left[\frac{(\lambda_0 - \lambda_1) + \Phi^{-1}(1 - \alpha)\eta_0/\sqrt{n_c}}{\eta_1/\sqrt{n_c}}\right], \qquad (4.1)$$

L. Lin et al., *Statistical Tools for Measuring Agreement*,
DOI 10.1007/978-1-4614-0562-7_4, © Springer Science+Business Media, LLC 2012

and the power based on MSD becomes

$$P = \Phi\left[\frac{(\lambda_0 - \lambda_1) - \Phi^{-1}(1 - \alpha)\eta_0/\sqrt{n_c}}{\eta_1/\sqrt{n_c}}\right], \qquad (4.2)$$

where $\Phi(\cdot)$ is the cumulative standard normal density function. For the one-tailed fixed type-I and type-II errors α and β, the associated sample size becomes

$$n = \left[\frac{\Phi^{-1}(1 - \beta)\eta_1 + \Phi^{-1}(1 - \alpha)\eta_0}{(\lambda_0 - \lambda_1)}\right]^2 + c. \qquad (4.3)$$

4.2 The Simplified Case

For the case of two raters with a single reading per rater, we can simplify the computations of the power and sample size using the upper bound of the variance of each estimate of the agreement indices. According to (2.29) and (2.30), the variance of the Z-transformed CCC estimate is less than or equal to $1/(n-2)$. Such approximation is almost exact when υ is close to zero and ϖ is close to one. According to (2.3) and (2.4), the variance of the log-transformed MSD estimate is less than or equal to $2/(n-2)$. Such approximation is exact when υ is equal to zero. The upper bound of the variance of transformed CP estimate remains unknown. Because CP is a mirrored index of TDI, and the approximated CP can also be computed from MSD, we need only calculate the sample size of approximated TDI or CP, which can be derived from the upper bound of the variance of the log-transformed MSD estimate. The sample size determination does not have to be exact, as long as we stay on the conservative side. For categorical data, even though the Z-transformation does not necessarily help, we can still utilize the simplification for computing the sample size for the weighted kappa, since the variances of CCC and weighted kappa are the same under the GEE methodology.

4.3 Examples Based on the Simplified Case

Suppose, as an example, that the available historical data indicate that CCC is 0.99 over a desirable data range, and we are willing to tolerate a CCC of 0.98. In this case, the sample size would be

$$n = \left[\frac{\Phi^{-1}(1 - \beta) + \Phi^{-1}(1 - \alpha)}{\tanh^{-1}(0.99) - \tanh^{-1}(0.98)}\right]^2 + 2,$$

which is equal to $n = 53$ for $\alpha = 0.05$ and $1 - \beta = 0.8$, where $\tanh^{-1}(\cdot)$ is the Z-transformation.

Suppose, as another example, the historical data indicate that $\text{TDI}\%_{0.9}$ is 10%, and we are willing to tolerate a $\text{TDI}\%_{0.9}$ of 15%. According to (2.42), the relationship between TDI based on the log-transformed data and TDI% is $\text{TDI} = \ln(1 + \frac{\text{TDI}\%}{100})$. In addition, MSD is proportional to TDI^2. In this case, the proportionality value between MSD and TDI^2 is irrelevant, and the sample size would become

$$n = 2\left(\frac{\Phi^{-1}(1-\beta) + \Phi^{-1}(1-\alpha)}{\ln\left\{[\ln(1+0.15)]^2/[\ln(1+0.10)]^2\right\}}\right)^2 + 2,$$

which is equal to $n = 24$ for $\alpha = 0.05$ and $1 - \beta = 0.8$, where $\ln(\cdot)$ is the log transformation.

Most frequently, we only have historical within sample variance (σ_e^2) for data with constant error or coefficient of variation (CV%) for data with proportional error. We can easily translate these into TDI and CCC. Recall that in (2.24), CCC is inversely related to the mean square of the ratio of the within-sample total deviation and the total deviation. When the marginal distributions of two raters are identical, the within-sample total deviation is σ_e and the total deviation is the square root of the sum of σ_e^2 and the between-sample variance. The data range is proportional to the associated standard deviation, and assuming that our historical σ_e was computed based on a large sample size (say at least 100 observations), this proportionality value is about 5 (Grant and Leavenworth 1972).

For data with constant error, we can use $\Phi^{-1}[1-(1-\pi)/2]\sqrt{2}\sigma_e$ as our historical TDI_{π_0} value, and use $1 - (5\sigma_e^2)/d_r^2$ as our historical CCC value over a desirable data range d_r. For data with proportional error, we can assume that the data are log-normally distributed and use the relationship

$$\omega^2 = \exp(\sigma_e^2) - 1, \tag{4.4}$$

where σ_e^2 is the historical within-sample variance based on log-transformed data and ω is the coefficient of variation (CV). As for the allowable TDI and CCC, we simply allow a cushion for σ_e^2 and some error for the biased square of the two raters, σ_β^2.

For example, if our historical CV is 10%, we have our historical σ_e^2 of $\ln(1 + 0.1^2) = 0.01$ based on log-transformed data, or $\text{TDI}_{0.9}$ of $1.645\sqrt{2\ln(1 + 0.1^2)} = 0.232$, or $\text{TDI}\%_{0.9}$ of 26.1%. Assuming that the maximum value of the data is ten times the minimum value, we have a historical CCC of $1 - 0.01/[\ln(10)/5]^2 = 0.953$. If we are willing to allow for a 50% increase in σ_e^2, and a σ_β^2 that is half of σ_e^2, we have an allowable CV% of $\sqrt{\exp(0.01 \times 1.5 + 0.01 \times 0.5) - 1} = 14.2\%$, or $\text{TDI}_{0.9}$ of 0.328, or $\text{TDI}\%_{0.9}$ of 38.5%, and an allowable CCC of 0.906. In this case, the sample size based on TDI would be

$$n = 2\left[\frac{\Phi^{-1}(1-\beta) + \Phi^{-1}(1-\alpha)}{\ln(0.328^2) - \ln(0.232^2)}\right]^2 + 2,$$

which is equal to $n = 28$ for $\alpha = 0.05$ and $1 - \beta = 0.8$. The sample size based on CCC would be

$$n = \left[\frac{\Phi^{-1}(1 - \beta) + \Phi^{-1}(1 - \alpha)}{\tanh^{-1}(0.953) - \tanh^{-1}(0.906)} \right]^2 + 2,$$

which is equal to $n = 51$ for $\alpha = 0.05$ and $1 - \beta = 0.8$. Here, the value of d_r is relatively irrelevant in computing the sample size. Note that the historical $\text{TDI}_{0.9}$ and CCC calculated from CV% actually correspond to the intrasample TDI and CCC. We need to have a little more cushion to allow for some systematic bias and for more random error when setting the criterion.

As stated earlier, the sample size (power) based on CCC is always larger (smaller) than that based on TDI, primarily due to the additional variation related to estimating the scaled denominator. After collecting the data, we will accept the agreement if the one-sided 95% upper (lower) confidence limit of the TDI (CCC) is better than the preset criterion.

Chapter 5
A Unified Model for Continuous and Categorical Data

In this chapter, we generalize agreement assessment for continuous and categorical data to cover multiple raters ($k \geq 2$), and each with multiple readings ($m \geq 1$) from each of the n subjects. In Chapters 2 and 3, we discussed agreement statistics for continuous and categorical data, respectively, based on the basic model of two raters with a single measure each per subject. In the terminology of this chapter, those earlier chapters discussed primarily the case $k = 2$ and $m = 1$. We utilize the results from Barnhart, Song, and Haber (2005), who first proposed the within-rater CCC, between-rater CCC based on the average of replicates, and between-rater CCC based on individual replicate, and used GEE methodology for estimation and statistical inference. We then combine the GEE methodology with the knowledge gained from Robieson (1999) and Carrasco and Jover (2003), and propose a unified approach that is applicable to continuous and categorical data.

Our approach establishes the agreement statistics of CCC, accuracy and precision coefficients for continuous and categorical data, and TDI and CP for normally distributed data, based on functions of variance components through a two-way mixed-effect model. We segregate these agreement statistics of MSD, TDI, CP, CCC, precision and accuracy coefficients into intrarater, interrater based on the average of replicates, and total rater based on individual replicate values. We then use the GEE methodology to form the estimations and combine it with the delta method to form statistical inferences. This chapter is largely based on the materials from Lin, Hedayat, and Wu (2007).

Suppose each of k raters measures each of n subjects m times. The model we use for measuring agreement is

$$y_{ijl} = \mu + \alpha_i + \beta_j + \gamma_{ij} + e_{ijl}, \tag{5.1}$$

where y_{ijl} stands for the lth reading from subject i given by rater j, with $i = 1, 2, \ldots, n$, $j = 1, 2, \ldots, k$ and $l = 1, 2, \ldots, m$. The readings can be continuous, binary, or ordinal. The overall mean is μ. The random subject effect, α_i, has equal second moments across all raters. The random rater and subject interaction

L. Lin et al., *Statistical Tools for Measuring Agreement*,
DOI 10.1007/978-1-4614-0562-7_5, © Springer Science+Business Media, LLC 2012

effect, γ_{ij}, has equal second moments across all raters. The random error effect, e_{ijl}, is uncorrelated with α_i and γ_{ij}. The fixed rater effect is β_j, and we assume $\sum_{j=1}^{k} \beta_j = 0$.

The random subject effect α_i has mean 0 and variance σ_α^2. The interaction effect γ_{ij} has mean 0 and variance σ_γ^2. The random error effect has mean 0 and variance σ_e^2.

5.1　Definition of Variance Components

Based on model (5.1) and balanced data, variance components can be expressed as follows. First, the variance of the random error effect is defined as

$$\sigma_e^2 = \frac{\sum_{i=1}^{n} \sum_{j=1}^{k} \sigma_{ij}^2}{nk}, \tag{5.2}$$

where σ_{ij}^2 is the variance of y_{ijl} for subject i and rater j.

The variance of the random subject effect is defined as

$$\sigma_\alpha^2 = \frac{4 \sum_{j=1}^{k-1} \sum_{j'=j+1}^{k} \sum_{l=1}^{m-1} \sum_{l'=l+1}^{m} \sigma_{jj'll'}}{m^2 k(k-1)}, \tag{5.3}$$

where $\sigma_{jj'll'}$ is the covariance of y_{ijl} and $y_{ij'l'}$ among different raters and replicates.

The variance of the interaction effect is defined as

$$\sigma_\gamma^2 = A + B - C - D, \tag{5.4}$$

where

$$A = \frac{\sum_{j=1}^{k} \sum_{l=1}^{m} \sigma_{jl}^2}{m^2 k}, \tag{5.5}$$

and σ_{jl}^2 is the variance of y_{ijl} for rater j and replicate l,

$$B = \frac{2 \sum_{j=1}^{k} \sum_{l=1}^{m-1} \sum_{l'=l+1}^{m} \sigma_{jll'}}{m^2 k}, \tag{5.6}$$

where $\sigma_{jll'}$ is the covariance of y_{ijl} and $y_{ijl'}$ between replicates for rater j,

$$C = \sigma_\alpha^2, \tag{5.7}$$

and

$$D = \frac{\sigma_e^2}{m}. \tag{5.8}$$

Finally, the mean square of the fixed rater effect is defined as

$$\sigma_\beta^2 = \frac{\sum_{j=1}^{k-1} \sum_{j'=j+1}^{k} (\beta_j - \beta_{j'})^2}{k(k-1)}. \tag{5.9}$$

5.2 Intrarater Precision

We assume that replicates within a rater are interchangeable. When measuring unscaled intrarater agreement independent of data range, we use ε_{intra}^2 to denote the MSD_{intra} between any two replications l and l', $l, l' = 1, 2, \ldots, m$, for any raters or for the average across k fixed raters:

$$\begin{aligned}
\varepsilon_{intra}^2 &= E(y_{ijl} - y_{ijl'})^2 \\
&= (\mu_j - \mu_j)^2 + 2\left(\sigma_\alpha^2 + \sigma_\gamma^2 + \sigma_e^2\right) - 2(\sigma_\alpha^2 + \sigma_\gamma^2) \\
&\quad - 2\sigma_e^2.
\end{aligned} \tag{5.10}$$

For normally distributed data, following (2.8), $\mathrm{TDI}_{intra(\pi_0)}$ can be expressed as

$$\delta_{intra(\pi_0)} = \Phi^{-1}\left(1 - \frac{1-\pi_0}{2}\right)\sqrt{2\sigma_e^2}, \tag{5.11}$$

and $\mathrm{CP}_{intra(\delta_0)}$ can be expressed as

$$\pi_{intra(\delta_0)} = 1 - 2\left[1 - \Phi\left(\frac{\delta_0}{\sqrt{2\sigma_e^2}}\right)\right]. \tag{5.12}$$

For any rater j, the intrarater precision for continuous and categorical data between any two replications, l and l', is defined as

$$\rho_{intra} = \frac{\mathrm{cov}(y_{ijl}, y_{ijl'})}{\sqrt{\mathrm{var}(y_{ijl})}\sqrt{\mathrm{var}(y_{ijl'})}} = 1 - \frac{\varepsilon_{intra}^2}{\varepsilon_{intra}^2|_{\rho_{yij1,yij2,\ldots,yijm}=0}}, \tag{5.13}$$

where $\mathrm{var}(\cdot)$ and $\mathrm{cov}(\cdot)$ represent the variance and covariance functions, respectively.

Under model (5.1) in terms of variance components defined in Section 5.1, the CCC_{intra} becomes

$$\rho_{c,intra} = \rho_{intra} = \frac{\sigma_\alpha^2 + \sigma_\gamma^2}{\sigma_\alpha^2 + \sigma_\gamma^2 + \sigma_e^2}. \tag{5.14}$$

The CCC$_{\text{intra}}$ measures the proportion of the variance that is attributable to the subjects. Based on model (5.1), this proportion is the same for all k raters. Furthermore, the means of replicates within a rater are assumed equal under model (5.1). Therefore, the intrarater agreement CCC$_{\text{intra}}$ equals ρ_{intra}, the intrarater precision coefficient with the accuracy coefficient of one, and (5.11) and (5.12) are exact, not approximate. This relative agreement index is heavily dependent on the total variability (total data range).

5.3   Interrater Agreement

Since there are m replicated readings for subject i given by rater j, the average of those m readings could be used to measure the interrater agreement. We use $\bar{y}_{ij\cdot}$ to denote the average of m readings from subject i given by rater j.

The MSD between any two raters j and j', MSD$_{\text{inter}_{jj'}}$, becomes

$$
\varepsilon^2_{\text{inter}_{jj'}} = E\left[(\bar{y}_{ij\cdot} - \bar{y}_{ij'\cdot})^2\right]
$$

$$
= (\beta_j - \beta_{j'})^2 + 2\left(\sigma^2_\gamma + \frac{\sigma^2_e}{m}\right). \tag{5.15}
$$

Across k fixed raters, MSD$_{\text{inter}}$ is the average of $k(k-1)/2$ MSD$_{\text{inter}_{jj'}}$ indices

$$
\varepsilon^2_{\text{inter}} = \frac{2}{k(k-1)} \sum_{j=1}^{k-1} \sum_{j'=j+1}^{k} \varepsilon^2_{\text{inter}_{jj'}}
$$

$$
= 2\left(\sigma^2_\beta + \sigma^2_\gamma + \frac{\sigma^2_e}{m}\right). \tag{5.16}
$$

In this chapter we use the approximated inter and total CP based on (2.22) because the exact CP based on (2.10) would be complicated for model (5.1). For normally distributed data, TDI$_{\text{inter}(\pi_0)}$ and CP$_{\text{inter}(\delta_0)}$ can be approximated by

$$
\delta_{\text{inter}(\pi_0)} \doteq \Phi^{-1}\left(1 - \frac{1-\pi_0}{2}\right) \sqrt{2\sigma^2_\beta + 2\sigma^2_\gamma + 2\frac{\sigma^2_e}{m}} \tag{5.17}
$$

and

$$
\pi_{\text{inter}(\delta_0)} \doteq 1 - 2\left[1 - \Phi\left(\frac{\delta_0}{\sqrt{2\sigma^2_\beta + 2\sigma^2_\gamma + 2\sigma^2_e/m}}\right)\right]. \tag{5.18}
$$

The TDI and CP approximations are good when the relative biased squared (RBS) value is reasonable (see Section 2.2.2). Otherwise, the approximation will be conservative when $\pi_0 \geq 0.9$. Here, the RBS is defined as:

$$\Delta_{\text{inter}} = \frac{\sigma_\beta^2}{\sigma_\gamma^2 + \sigma_e^2/m}. \tag{5.19}$$

For continuous and categorical data, the $\text{CCC}_{\text{inter}}$ becomes

$$\rho_{c,\text{inter}} = 1 - \frac{\varepsilon_{\text{inter}}^2}{\varepsilon_{\text{inter}}^2|_{\rho_{\bar{y}_{i1}\cdot\cdot\bar{y}_{i2}\cdots\cdots\bar{y}_{ik\cdot}}=0}}. \tag{5.20}$$

Under model (5.1) in terms of variance components defined in Section 5.1, the $\text{CCC}_{\text{inter}}$ becomes

$$\rho_{c,\text{inter}} = \frac{\sigma_\alpha^2}{\sigma_\alpha^2 + \sigma_\gamma^2 + \frac{\sigma_e^2}{m} + \sigma_\beta^2}. \tag{5.21}$$

Since readings from different raters have different expected means, we further define that the interrater agreement $\text{CCC}_{\text{inter}}$ consists of two parts: interrater precision and interrater accuracy coefficients. The interrater precision coefficient becomes

$$\rho_{\text{inter}} = \frac{\text{cov}(\bar{y}_{ij\cdot}, \bar{y}_{ij'\cdot})}{\sqrt{\text{var}(\bar{y}_{ij\cdot})}\sqrt{\text{var}(\bar{y}_{ij'\cdot})}} = \frac{\sigma_\alpha^2}{\sigma_\alpha^2 + \sigma_\gamma^2 + \frac{\sigma_e^2}{m}}. \tag{5.22}$$

The interrater accuracy coefficient becomes

$$\chi_{a,\text{inter}} = \frac{\sigma_\alpha^2 + \sigma_\gamma^2 + \frac{\sigma_e^2}{m}}{\sigma_\alpha^2 + \sigma_\gamma^2 + \frac{\sigma_e^2}{m} + \sigma_\beta^2}. \tag{5.23}$$

Here, $\rho_{c,\text{inter}}$ is the product of ρ_{inter} and $\chi_{a,\text{inter}}$. The accuracy index measures how close the means of raters are. In model (5.1), variances are assumed to be the same for different raters, and consequently they are not present in the accuracy index. Therefore, the definition of accuracy is slightly modified compared to that originally defined by Lin (1989). The interrater agreement is measured based on the average of m readings made by each rater. Therefore, the agreement indices depend on the number of replications (m).

The approach proposed by Barnhart, Song, and Haber (2005) allows for different variances among raters, and is a measure based on the true readings, μ_{ij}, from each rater and subject. Therefore, their interrater CCC does not depend on the number of replications. In addition, the interrater CCC from Barnhart, Song, and Haber (2005) equals the limit of our $\text{CCC}_{\text{inter}}$ as the number of replications m goes to infinity.

5.4   Total-Rater Agreement

Since there are m replicated readings for subject i given by rater j, the interrater agreement could be based on any one of the m replicated readings. Total agreement is such a measure of agreement based on any individual reading from each rater. The MSD between any two raters j and j', $\text{MSD}_{\text{total}_{jj'}}$, becomes

$$
\varepsilon^2_{\text{total}_{jj'}} = E\left[(y_{ijl} - y_{ij'l'})^2\right]
$$

$$
= (\beta_j - \beta_{j'})^2 + 2\left(\sigma^2_\gamma + \sigma^2_e\right). \tag{5.24}
$$

Across k fixed raters, $\text{MSD}_{\text{total}}$ is the average of $k(k-1)/2\,\text{MSD}_{\text{total}jj'}$ indices

$$
\varepsilon^2_{\text{total}} = \frac{2}{k(k-1)} \sum_{j=1}^{k-1} \sum_{j'=j+1}^{k} \varepsilon^2_{\text{total}_{jj'}}
$$

$$
= 2\left(\sigma^2_\beta + \sigma^2_\gamma + \sigma^2_e\right). \tag{5.25}
$$

For normally distributed data, the $\text{TDI}_{\text{total}(\pi_0)}$ and $\text{CP}_{\text{total}(\delta_0)}$ can be approximated by

$$
\delta_{\text{total}(\pi_0)} \doteq \Phi^{-1}\left(1 - \frac{1-\pi_0}{2}\right)\sqrt{2\sigma^2_\beta + 2\sigma^2_\gamma + 2\sigma^2_e} \tag{5.26}
$$

and

$$
\pi_{\text{total}(\delta_0)} \doteq 1 - 2\left[1 - \Phi\left(\frac{\delta_0}{\sqrt{2\sigma^2_\beta + 2\sigma^2_\gamma + 2\sigma^2_e}}\right)\right]. \tag{5.27}
$$

The TDI and CP approximations are adequate when the RBS is reasonable. Otherwise, the approximations will be conservative when $\pi_0 \geq 0.9$. Here, the RBS is defined as

$$
\Delta_{\text{total}} = \frac{\sigma^2_\beta}{\sigma^2_\gamma + \sigma^2_e}. \tag{5.28}
$$

For continuous and categorical data, $\text{CCC}_{\text{total}}$ is defined as

$$
\rho_{c,\text{total}} = 1 - \frac{\varepsilon^2_{\text{total}}}{\varepsilon^2_{\text{total}|\rho_{y_{i1l}\cdot y_{i2l}\cdots y_{ikl}}=0}}. \tag{5.29}
$$

Under model (5.1) in terms of variance components defined in Section 5.1, the $\text{CCC}_{\text{total}}$ becomes

$$
\rho_{c,\text{total}} = \frac{\sigma^2_\alpha}{\sigma^2_\alpha + \sigma^2_\gamma + \sigma^2_e + \sigma^2_\beta}. \tag{5.30}
$$

The total-rater precision and accuracy coefficients are

$$\rho_{\text{total}} = \frac{\text{cov}(y_{ijl}, y_{ij'l'})}{\sqrt{\text{var}(y_{ijl})}\sqrt{\text{var}(y_{ij'l'})}}$$

$$= \frac{\sigma_\alpha^2}{\sigma_\alpha^2 + \sigma_\gamma^2 + \sigma_e^2} \tag{5.31}$$

and

$$\chi_{a,\text{total}} = \frac{\sigma_\alpha^2 + \sigma_\gamma^2 + \sigma_e^2}{\sigma_\alpha^2 + \sigma_\gamma^2 + \sigma_e^2 + \sigma_\beta^2}. \tag{5.32}$$

Again,

$$\rho_{c,\text{total}} = \rho_{\text{total}} \chi_{a,\text{total}}.$$

5.5   Proportional Error Case

Similar to Section 2.5, when the residual standard deviation becomes proportional to the measurement, we apply the natural log transformation to the data and then compute the agreement statistics. The TDI%$_{\pi_0}$ is defined as

$$\theta_{\pi_0} = 100 \left[\exp(\delta_{\pi_0}) - 1\right] \%. \tag{5.33}$$

In this situation, TDI%$_{\pi_0}$ is the antitransformed TDI$_{\pi_0}$ from which 1 is subtracted, which measures a percent change rather than an absolute deviation.

5.6   Asymptotic Normality

We use $\bar{y}_{.jl}$ to denote the average of n readings from rater j given by replicate l. The rater means and all variance components, $\mu_1, \mu_2, \ldots, \mu_k, \sigma_\beta^2, \sigma_\alpha^2, \sigma_\gamma^2, \sigma_e^2$ are estimated through the GEE methodology according to the following system of equations:

$$\sum_{i=1}^{n} F_i' H_i^{-1}[Q_i - \Theta] = 0. \tag{5.34}$$

Here,

$$
Q_i =
\begin{pmatrix}
\frac{1}{m}(y_{i11} + y_{i12} + \cdots + y_{i1m}) \\
\vdots \\
\frac{1}{m}(y_{ij1} + y_{ij2} + \cdots + y_{ijm}) \\
\vdots \\
\frac{1}{m}(y_{ik1} + y_{ik2} + \cdots + y_{ikm}) \\
\frac{1}{k(k-1)} \sum_{j=1}^{k-1} \sum_{j'=j+1}^{k} (\bar{y}_{ij\cdot} - \bar{y}_{ij'\cdot})^2 \\
\frac{4}{m^2 k(k-1)} \sum_{j=1}^{k-1} \sum_{j'=j+1}^{k} \sum_{l=1}^{m-1} \sum_{l'=l+1}^{m} \left[(y_{ijl} - \bar{y}_{\cdot jl})(y_{ij'l'} - \bar{y}_{\cdot j'l'}) \right] \\
\frac{1}{k} \sum_{j=1}^{k} \left[\frac{\sum_{l=1}^{m}(y_{ijl} - \bar{y}_{ij\cdot})^2}{(m-1)} \right] \\
\frac{1}{m^2 k} \sum_{j=1}^{k} \sum_{l=1}^{m}(y_{ijl} - \bar{y}_{\cdot jl})^2 + \frac{2}{m^2 k} \sum_{j=1}^{k} \sum_{l=1}^{m-1} \sum_{l'=l+1}^{m}(y_{ijl} - \bar{y}_{\cdot jl})(y_{ijl'} - \bar{y}_{\cdot jl'})
\end{pmatrix},
$$

with the expected values of

$$
\Theta =
\begin{pmatrix}
\mu_1 \\
\vdots \\
\mu_j \\
\vdots \\
\mu_k \\
\sigma_\beta^2 + \frac{\sigma_e^2}{m} + \sigma_\gamma^2 \\
\sigma_\alpha^2 \\
\sigma_e^2 \\
\sigma_\alpha^2 + \frac{\sigma_e^2}{m} + \sigma_\gamma^2
\end{pmatrix}.
$$

The working covariance matrix for Q_i (Zeger and Liang 1986) is conveniently set as a diagonal matrix (Barnhart and Williamson 2001) given by

$$H_i = \text{diag}(\text{var}(Q_i)) = \begin{pmatrix} a_1 & & & & & & \\ & \ddots & & & & \mathbf{0} & \\ & & a_j & & & & \\ & & & \ddots & & & \\ & & & & a_k & & \\ & & & & & d & \\ & \mathbf{0} & & & & & e \\ & & & & & & & f \\ & & & & & & & & g \end{pmatrix},$$

with the elements of

$$a_1 = a_j = a_k = \text{var}\left[\frac{1}{m}(y_{ij1} + y_{ij2} + \cdots + y_{ijm})\right]$$

$$= \sigma_\alpha^2 + \sigma_\gamma^2 + \frac{\sigma_e^2}{m}, \tag{5.35}$$

$$d = \text{var}\left[\frac{1}{k(k-1)} \sum_{j=1}^{k-1} \sum_{j'=j+1}^{k} (\bar{y}_{ij.} - \bar{y}_{ij'.})^2\right]$$

$$= \frac{k(k-1)(3k-2)}{m^2}\sigma_e^4 + k(k-1)(3k-2)\sigma_\gamma^4 + 8k(k-1)\sigma_\beta^2\sigma_\gamma^2$$

$$+ \frac{8k(k-1)}{m}\sigma_\beta^2\sigma_e^2 + \frac{4k^2(k-1)}{m}\sigma_\gamma^2\sigma_e^2, \tag{5.36}$$

$$e = \text{var}\left\{\frac{4}{m^2k(k-1)} \sum_{j=1}^{k-1} \sum_{j'=j+1}^{k} \sum_{l=1}^{m-1} \sum_{l'=l+1}^{m} \left[(y_{ijl} - \bar{y}_{.jl})(y_{ij'l'} - \bar{y}_{.j'l'})\right]\right\}$$

$$= m^4 k(k-1)(2k-3)\sigma_\alpha^4 + \frac{m^4 k(k-1)(2k-3)}{2}\sigma_\gamma^4 + \frac{m^2 k(k-1)}{2}\sigma_e^4$$

$$+ \frac{\left[2 + (m-1)^2 + 2m(m-1)(k-2)\right]k(k-1)m^2}{2}\left(\sigma_\alpha^2\sigma_e^2 + \sigma_\gamma^2\sigma_e^2\right)$$

$$+ m^4 k(k-1)(2k-3)\sigma_\alpha^2\sigma_\gamma^2, \tag{5.37}$$

$$f = \text{var}\left\{\frac{1}{k} \sum_{j=1}^{k}\left[\frac{\sum_{l=1}^{m}(y_{ijl} - \bar{y}_{ij.})^2}{(m-1)}\right]\right\}$$

$$= \frac{2k}{m-1}\sigma_e^4, \tag{5.38}$$

and

$$g =$$

$$\text{var}\left[\frac{1}{m^2 k}\sum_{j=1}^{k}\sum_{l=1}^{m}(y_{ijl}-\bar{y}_{\cdot jl})^2 + \frac{2}{m^2 k}\sum_{j=1}^{k}\sum_{l=1}^{m-1}\sum_{l'=l+1}^{m}(y_{ijl}-\bar{y}_{\cdot jl})(y_{ijl'}-\bar{y}_{\cdot jl'})\right]$$

$$= \left[km(m-1)(2m-3) + m(m-1)(k-1)/2 + \frac{m^2 k^2(3m-1)}{2}\right]\sigma_\alpha^4$$

$$+ \left[m^2 k\left(1-\frac{3k}{2}+m+\frac{mk}{2}\right) + km(m-1)(2m-3)\right]\sigma_\gamma^4 + \frac{mk(m-3)}{2}\sigma_e^4$$

$$+ \left[2m^2 k(2-k) + mk(m-1)(mk+5m-4)\right]\sigma_\alpha^2\sigma_\gamma^2$$

$$+ \left[mk(4-mk+(m-1)^2 + \frac{mk(m-1)}{2}\right]\sigma_\alpha^2\sigma_e^2$$

$$+ mk\left[4+(m-1)^2-mk+\frac{(m-1)(mk+4)}{2}\right]\sigma_\gamma^2\sigma_e^2. \tag{5.39}$$

Finally,

$$F_i = \frac{\partial\Theta}{\partial(\mu_1,\ldots,\mu_k,\sigma_\beta^2,\sigma_\alpha^2,\sigma_e^2,\sigma_\gamma^2)} = \begin{pmatrix} \mathbf{1}_k & \mathbf{0}_{4\times 4} \\ \mathbf{0}_{4\times 4} & \boldsymbol{f}_{4\times 4} \end{pmatrix}, \tag{5.40}$$

where

$$\boldsymbol{f}_{4\times 4} = \begin{pmatrix} 1 & 0 & 1/m & 1 \\ 0 & 1 & 0 & 0 \\ 0 & 0 & 1 & 0 \\ 0 & 1 & 1/m & 1 \end{pmatrix}.$$

Note that the working covariance matrix is derived assuming normality. We obtain the estimates of the means and variance components through (5.34), which are their respective sample counterparts given in $\boldsymbol{Q}_i$, and estimate their variance–covariance matrix through the GEE methodology.

The variance–covariance matrix for $\boldsymbol{Q}_i$ is

$$\text{var}(\boldsymbol{Q}_i) = \frac{1}{n}\boldsymbol{D}^{-1'}\boldsymbol{\Sigma}\boldsymbol{D}^{-1}, \tag{5.41}$$

where

$$\boldsymbol{D} = \sum_{i=1}^{n}\boldsymbol{F}_i'\boldsymbol{H}_i^{-1}\boldsymbol{F}_i$$

and

$$\Sigma = \sum_{i=1}^{n} F_i' H_i^{-1} (Q_i - \Theta)(Q_i - \Theta)' H_i^{-1} F_i.$$

Since the working covariance matrix, H_i, in (5.34) is a diagonal matrix and model (5.1) does not involve covariates, the estimate of $\text{var}(Q_i)$ actually degenerates to the robust variance estimate from the least squares estimation equations (Zeger, Liang, and Albert 1988).

We then use the delta method to derive the asymptotic variance of the estimate for each agreement index. We arrive analytically at the following variances of the MSD estimates:

$$\text{var}(\hat{\varepsilon}_{\text{intra}}^2) = \frac{4}{n} \text{var}(\sigma_e^2), \tag{5.42}$$

$$\text{var}(\hat{\varepsilon}_{\text{inter}}^2) = \frac{4}{n} \Bigg[\text{var}(\sigma_\beta^2) + \text{var}(\sigma_\gamma^2) + \frac{\text{var}(\sigma_e^2)}{m^2}$$

$$+ 2\text{cov}(\sigma_\beta^2, \sigma_\gamma^2) + \frac{2\text{cov}(\sigma_\beta^2, \sigma_e^2)}{m} + \frac{2\text{cov}(\sigma_e^2, \sigma_\gamma^2)}{m} \Bigg], \tag{5.43}$$

and

$$\text{var}(\hat{\varepsilon}_{\text{total}}^2) = \frac{4}{n} \Bigg[\text{var}(\sigma_\beta^2) + \text{var}(\sigma_\gamma^2) + \text{var}(\sigma_e^2)$$

$$+ 2\text{cov}(\sigma_\beta^2, \sigma_\gamma^2) + 2\text{cov}(\sigma_\beta^2, \sigma_e^2) + 2\text{cov}(\sigma_e^2, \sigma_\gamma^2) \Bigg]. \tag{5.44}$$

When estimating the variances of TDIs, we use the log transformation based on MSD, $W_{(\cdot)} = \ln(\varepsilon_{(\cdot)}^2)$. The transformed variance for w is

$$\text{var}(W_{(\cdot)}) = \frac{\text{var}(\varepsilon_{(\cdot)}^2)}{\varepsilon_{(\cdot)}^4}.$$

Therefore, we have

$$\text{var}(\widehat{W}_{\text{intra}}) = \frac{\text{var}(\sigma_e^2)}{n(\sigma_e^2)^2}, \tag{5.45}$$

$$\text{var}(\widehat{W}_{\text{inter}}) = \frac{\text{var}(\sigma_\beta^2) + \text{var}(\sigma_\gamma^2) + \text{var}(\sigma_e^2)/m^2}{n(\sigma_\beta^2 + \sigma_\gamma^2 + \sigma_e^2/m)^2}$$

$$+ \frac{\text{cov}(\sigma_\beta^2, \sigma_\gamma^2) + \text{cov}(\sigma_\beta^2, \sigma_e^2)/m + \text{cov}(\sigma_e^2, \sigma_\gamma^2)/m}{n(\sigma_\beta^2 + \sigma_\gamma^2 + \sigma_e^2/m)^2}, \tag{5.46}$$

and

$$\text{var}(\widehat{W}_{\text{total}}) = \frac{\text{var}(\sigma_\beta^2) + \text{var}(\sigma_\gamma^2) + \text{var}(\sigma_e^2)}{n(\sigma_\beta^2 + \sigma_\gamma^2 + \sigma_e^2)^2}$$

$$+ \frac{\text{cov}(\sigma_\beta^2, \sigma_\gamma^2) + \text{cov}(\sigma_\beta^2, \sigma_e^2) + \text{cov}(\sigma_e^2, \sigma_\gamma^2)}{n(\sigma_\beta^2 + \sigma_\gamma^2 + \sigma_e^2)^2}. \tag{5.47}$$

The variances of $\text{TDI}_{\text{intra}}$, $\text{TDI}_{\text{inter}}$, and $\text{TDI}_{\text{total}}$ are given in (5.45), (5.46), and (5.47) divided by 4, respectively.

For CP indices, we use the asymptotic variance based on the transformed variable from (5.12), (5.18), and (5.27), and we have

$$\text{var}(\hat{\pi}_{(\cdot)\delta_0}) = \frac{e^{-\frac{\delta_0^2}{\varepsilon_{(\cdot)}^2}}}{n} \left(1 + \frac{\delta_0^2}{\varepsilon_{(\cdot)}^2}\right)^2 \frac{\text{var}(\varepsilon_{(\cdot)}^2)}{8\pi\varepsilon_{(\cdot)}^2\delta_0^2}, \tag{5.48}$$

where $\varepsilon_{(\cdot)}^2$ is intra, inter, or total MSD.

We arrive analytically at the following variances of the estimates of CCC, precision and accuracy coefficients:

$$\text{var}(\hat{\rho}_{c,\text{intra}}) = \text{var}(\hat{\rho}_{\text{intra}}) = \frac{(1 - \rho_{\text{intra}})^2}{n(\sigma_\alpha^2 + \sigma_\gamma^2 + \sigma_e^2)^2} \left\{ \left[\text{var}(\sigma_\alpha^2) + \text{var}(\sigma_\gamma^2) + 2\text{cov}(\sigma_\alpha^2, \sigma_\gamma^2)\right] \right.$$

$$\left. + \rho_{\text{intra}}^2 \text{var}(\sigma_e^2) - 2(1 - \rho_{\text{intra}})\rho_{\text{intra}}\left[\text{cov}(\sigma_\alpha^2, \sigma_e^2) + \text{cov}(\sigma_e^2, \sigma_\gamma^2)\right]\right\}, \tag{5.49}$$

$$\text{var}(\hat{\rho}_{c,\text{inter}}) = \frac{1}{n(\sigma_\alpha^2 + \sigma_\beta^2 + \sigma_\gamma^2 + \sigma_e^2/m)^2} \left\{(1 - \rho_{\text{inter}})^2\text{var}(\sigma_\alpha^2) + \rho_{\text{inter}}^2\left[\text{var}(\sigma_\beta^2)\right.\right.$$

$$\left. + \frac{\text{var}(\sigma_e^2)}{m^2} + \text{var}(\sigma_\gamma^2) + 2\text{cov}(\sigma_\beta^2, \sigma_\gamma^2) + \frac{2\text{cov}(\sigma_\beta^2, \sigma_e^2)}{m} + \frac{2\text{cov}(\sigma_e^2, \sigma_\gamma^2)}{m}\right]$$

$$\left. - 2(1 - \rho_{\text{inter}})\rho_{\text{inter}}\left[\text{cov}(\sigma_\beta^2, \sigma_\alpha^2) + \text{cov}(\sigma_\alpha^2, \sigma_\gamma^2) + \frac{\text{cov}(\sigma_\alpha^2, \sigma_e^2)}{m}\right]\right\}, \tag{5.50}$$

$$\text{var}(\hat{\rho}_{c,\text{total}}) = \frac{1}{n(\sigma_\alpha^2 + \sigma_\beta^2 + \sigma_\gamma^2 + \sigma_e^2)^2} \left\{(1 - \rho_{\text{total}})^2\text{var}(\sigma_\alpha^2) + \rho_{\text{total}}^2\left[\text{var}(\sigma_\beta^2) + \text{var}(\sigma_e^2)\right.\right.$$

$$\left. + \text{var}(\sigma_\gamma^2) + 2\text{cov}(\sigma_\beta^2, \sigma_\gamma^2) + 2\text{cov}(\sigma_\beta^2, \sigma_e^2) + 2\text{cov}(\sigma_e^2, \sigma_\gamma^2)\right]$$

$$\left. - 2(1 - \rho_{\text{total}})\rho_{\text{total}}\left[\text{cov}(\sigma_\beta^2, \sigma_\alpha^2) + \text{cov}(\sigma_\alpha^2, \sigma_\gamma^2) + \text{cov}(\sigma_\alpha^2, \sigma_e^2)\right]\right\}, \tag{5.51}$$

$$\text{var}(\hat{\rho}_{\text{inter}}) = \frac{1}{n(\sigma_\alpha^2 + \sigma_\gamma^2 + \sigma_e^2/m)^2} \left\{(1 - \rho_{\text{inter}})^2\text{var}(\sigma_\alpha^2) + \rho_{\text{inter}}^2\left[\frac{\text{var}(\sigma_e^2)}{m^2} + \text{var}(\sigma_\gamma^2)\right.\right.$$

$$\left. + \frac{2\text{cov}(\sigma_e^2, \sigma_\gamma^2)}{m}\right] - 2(1 - \rho_{\text{inter}})\rho_{\text{inter}}\left[\text{cov}(\sigma_\alpha^2, \sigma_\gamma^2) + \frac{\text{cov}(\sigma_\alpha^2, \sigma_e^2)}{m}\right]\right\}. \tag{5.52}$$

$$\text{var}(\hat{\rho}_{\text{total}}) = \frac{1}{n(\sigma_\alpha^2 + \sigma_\gamma^2 + \sigma_e^2)^2} \left\{ (1 - \rho_{\text{total}})^2 \text{var}(\sigma_\alpha^2) + \rho_{\text{total}}^2 \left[\text{var}(\sigma_e^2) + \text{var}(\sigma_\gamma^2) \right. \right.$$

$$\left. \left. + 2\text{cov}(\sigma_e^2, \sigma_\gamma^2) \right] - 2(1 - \rho_{\text{total}})\rho_{\text{total}} \left[\text{cov}(\sigma_\alpha^2, \sigma_\gamma^2) + \text{cov}(\sigma_\alpha^2, \sigma_e^2) \right] \right\}, \quad (5.53)$$

$$\text{var}(\hat{\chi}_{a,\text{inter}}) = \frac{1}{n(\sigma_\alpha^2 + \sigma_\beta^2 + \sigma_\gamma^2 + \sigma_e^2/m)^2} \left\{ (1 - \chi_{a,\text{inter}})^2 \left[\text{var}(\sigma_\alpha^2) + \text{var}(\sigma_\gamma^2) + \frac{\text{var}(\sigma_e^2)}{m^2} \right. \right.$$

$$\left. + \frac{2\text{cov}(\sigma_\alpha^2, \sigma_e^2)}{m} + 2\text{cov}(\sigma_\alpha^2, \sigma_\gamma^2) + \frac{2\text{cov}(\sigma_\gamma^2, \sigma_e^2)}{m} \right] + \chi_{a,\text{inter}}^2 \text{var}(\sigma_\beta^2)$$

$$\left. - 2(1 - \chi_{a,\text{inter}})\chi_{a,\text{inter}} \left[\text{cov}(\sigma_\alpha^2, \sigma_\beta^2) + \frac{\text{cov}(\sigma_\beta^2, \sigma_e^2)}{m} + \text{cov}(\sigma_\beta^2, \sigma_\gamma^2) \right] \right\}, \quad (5.54)$$

and

$$\text{var}(\hat{\chi}_{a,\text{total}}) = \frac{1}{n(\sigma_\alpha^2 + \sigma_\beta^2 + \sigma_\gamma^2 + \sigma_e^2)^2} \left\{ (1 - \chi_{a,\text{total}})^2 \left[\text{var}(\sigma_\alpha^2) + \text{var}(\sigma_\gamma^2) + \text{var}(\sigma_e^2) \right. \right.$$

$$\left. + 2\text{cov}(\sigma_\alpha^2, \sigma_e^2) + 2\text{cov}(\sigma_\alpha^2, \sigma_\gamma^2) + 2\text{cov}(\sigma_\gamma^2, \sigma_e^2) \right] + \chi_{a,\text{total}}^2 \text{var}(\sigma_\beta^2)$$

$$\left. - 2(1 - \chi_{a,\text{total}})\chi_{a,\text{total}} \left[\text{cov}(\sigma_\alpha^2, \sigma_\beta^2) + \text{cov}(\sigma_\beta^2, \sigma_e^2) + \text{cov}(\sigma_\beta^2, \sigma_\gamma^2) \right] \right\}. \quad (5.55)$$

When estimating the variances of the above CCC or precision coefficient estimates for continuous data, we use the Z-transformation. Thus the transformed variance of an estimate of CCC or precision indices is $\text{var}(Z_{\text{index}}) = \frac{\text{var}(\text{index})}{1 - (\text{index})^2}$, with the index being $\hat{\rho}_{c,\text{intra}}$, $\hat{\rho}_{c,\text{inter}}$, $\hat{\rho}_{c,\text{total}}$, $\hat{\rho}_{\text{inter}}$, or $\hat{\rho}_{\text{total}}$. When estimating the variances of accuracy coefficients or CP estimates for continuous data, we use the logit transformation. The transformed variance of an estimate of accuracy coefficient or CP is $\text{var}(\text{index}) = \frac{\text{var}(\text{index})}{(\text{index})(1 - \text{index})}$ with the index being $\hat{\chi}_{a,\text{inter}}$, $\hat{\chi}_{a,\text{total}}$, $\hat{\pi}_{\text{intra}(\delta_0)}$, $\hat{\pi}_{\text{inter}(\delta_0)}$, or $\hat{\pi}_{\text{total}(\delta_0)}$. When computing confidence limits of the above agreement indices for continuous data, we would compute the limit based on the respective transformation, and then antitransform the limit.

5.7 The Case $m = 1$

For the cases with $m = 1$, the interaction effect between rater and subject in model (5.1), γ_{ij}, cannot be separated from the error effect. Thus the model reduces to

$$y_{ij} = \mu + \alpha_i + \beta_j + e_{ij}, \quad (5.56)$$

where each term follows the same distribution as that specified in model (5.1). The variance components are simplified to

$$\sigma_\alpha^2 = \frac{2}{k(k-1)} \sum_{j=1}^{k-1} \sum_{j'=j+1}^{k} \sigma_{jj'}; \tag{5.57}$$

$$\sigma_e^2 = \frac{\sum_{j=j'=1}^{k} \sigma_{jj'}}{k} - \sigma_\alpha^2; \tag{5.58}$$

$$\sigma_\beta^2 = \frac{\sum_{j=1}^{k-1} \sum_{j'=j+1}^{k} (\mu_j - \mu_{j'})^2}{k(k-1)}. \tag{5.59}$$

Accordingly, the MSD, TDI, and CP for continuous data become

$$\varepsilon^2 = 2\sigma_e^2 + 2\sigma_\beta^2, \tag{5.60}$$

$$\delta_{\pi_0} \doteq \Phi^{-1}\left(1 - \frac{1-\pi_0}{2}\right) \sqrt{2\sigma_\beta^2 + 2\sigma_e^2}, \tag{5.61}$$

and

$$\pi_{\delta_0} \doteq 1 - 2\left[1 - \Phi\left(\frac{\delta_0}{\sqrt{2\sigma_\beta^2 + 2\sigma_e^2}}\right)\right]. \tag{5.62}$$

For continuous and categorical data, the CCC becomes

$$\begin{aligned}
\rho_c &= 1 - \frac{\varepsilon^2}{\varepsilon^2_{|\rho_{y_{i1},y_{i2},\dots,y_{ik}}=0}} \\[2mm]
&= \frac{\sigma_\alpha^2}{\sigma_\alpha^2 + \sigma_e^2 + \sigma_\beta^2} \\[2mm]
&= \frac{\sigma_\alpha^2}{\sigma_\alpha^2 + \sigma_e^2 + \frac{1}{k(k-1)} \sum_{j=1}^{k-1} \sum_{j'=j+1}^{k} (\mu_j - \mu_{j'})^2} \\[2mm]
&= \frac{\frac{2}{k(k-1)} \sum_{j=1}^{k-1} \sum_{j'=j+1}^{k} \sigma_{jj'}}{\frac{1}{k} \sum_{j=1}^{k} \sigma_j^2 + \frac{1}{k(k-1)} \sum_{j=1}^{k-1} \sum_{j'=j+1}^{k} (\mu_j - \mu_{j'})^2} \\[2mm]
&= \frac{2 \sum_{j=1}^{k-1} \sum_{j'=j+1}^{k} \sigma_{jj'}}{(k-1) \sum_{j=1}^{k} \sigma_j^2 + \sum_{j=1}^{k-1} \sum_{j'=j+1}^{k} (\mu_j - \mu_{j'})^2}. \tag{5.63}
\end{aligned}$$

The precision coefficient becomes

$$\rho = \frac{\sigma_\alpha^2}{\sigma_\alpha^2 + \sigma_e^2},$$

and the accuracy coefficient becomes

$$\chi_a = \frac{\sigma_\alpha^2 + \sigma_e^2}{\sigma_\alpha^2 + \sigma_\beta^2 + \sigma_e^2}.$$

The above equations show that each of the four variance components can be expressed as functions of variances and pairwise covariances. Thus even though in the proposed unified approach, we assume the homogeneity of all variances, the CCC defined in (5.63) remains the same as the overall concordance correlation coefficient (OCCC) proposed by Lin (1989), King and Chinchilli (2001a, 2001b), and Barnhart, Haber, and Song (2002), where they did not assume the homogeneity of all variances.

When we use the GEE methodology to estimate all variance components, the estimating system of equations given in (5.34) are simplified to

$$\boldsymbol{Q}_i = \begin{pmatrix} y_{i1} \\ \vdots \\ y_{ij} \\ \vdots \\ y_{ik} \\ \frac{1}{k(k-1)} \sum_{j=1}^{k-1} \sum_{j'=j+1}^{k} (y_{ij} - y_{ij'})^2 \\ \frac{1}{k} \sum_{j=1}^{k} (y_{ij} - \bar{y}_{\cdot j})^2 \\ \frac{2}{k(k-1)} \sum_{j=1}^{k} \sum_{j'=j+1}^{k} (y_{ij} - \bar{y}_{\cdot j})(y_{ijl'} - \bar{y}_{\cdot j'}) \end{pmatrix},$$

$$\boldsymbol{\Theta} = \begin{pmatrix} \mu_1 \\ \vdots \\ \mu_j \\ \vdots \\ \mu_k \\ \sigma_\beta^2 + \sigma_e^2 \\ \sigma_\alpha^2 + \sigma_e^2 \\ \sigma_\alpha^2 \end{pmatrix},$$

$$\boldsymbol{H}_i = \begin{pmatrix} (\sigma_\alpha^2 + \sigma_e^2)\boldsymbol{I}_k & 0 & 0 & 0 \\ 0 & \frac{\sigma_e^4 + \sigma_\beta^2 \sigma_e^2}{k(k-1)} & 0 & 0 \\ 0 & 0 & \frac{2k\sigma_\alpha^4 + \sigma_e^4 + 2\sigma_\alpha^2 \sigma_e^2}{k} & 0 \\ 0 & 0 & 0 & \frac{2k(k-1)\sigma_\alpha^4 + 2k\sigma_e^4 + 2\sigma_\alpha^2 \sigma_e^2}{k(k-1)} \end{pmatrix},$$

and

$$F_i = \frac{\partial \Theta}{\partial(\mu_1, \ldots, \mu_k, \sigma_\beta^2, \sigma_\alpha^2, \sigma_e^2)} = \begin{pmatrix} \mathbf{1}_k & 0 & 0 & 0 \\ 0 & 1 & 0 & 1 \\ 0 & 0 & 1 & 1 \\ 0 & 0 & 1 & 0 \end{pmatrix}.$$

The variances of the estimates of agreement indices are

$$\text{var}(\hat{\varepsilon}^2) = \frac{4}{n}\left[\text{var}(\sigma_\beta^2) + \text{var}(\sigma_e^2) + 2\text{cov}(\sigma_\beta^2, \sigma_e^2)\right], \tag{5.64}$$

$$\text{var}(\widehat{W}) = \frac{\text{var}(\hat{\varepsilon}^2)}{\varepsilon^4} = \frac{\text{var}(\sigma_\beta^2) + \text{var}(\sigma_e^2) + 2\text{cov}(\sigma_\beta^2, \sigma_e^2)}{n(\sigma_e^2 + \sigma_\beta^2)^2}, \tag{5.65}$$

$$\text{var}(\hat{\pi}_{\delta_0}) = \frac{e^{-\frac{\delta_0^2}{\varepsilon^2}}}{n}\left(1 + \frac{\delta_0^2}{\varepsilon^2}\right)^2 \frac{\text{var}(\varepsilon^2)}{8\pi\varepsilon^2\delta_0^2}, \tag{5.66}$$

$$\text{var}(\hat{\rho}_c) = \frac{1}{n(\sigma_\alpha^2 + \sigma_\beta^2 + \sigma_e^2)^2}\Big\{(1-\rho)^2\text{var}(\sigma_\alpha^2) + \rho^2[\text{var}(\sigma_\beta^2) + \text{var}(\sigma_e^2)$$

$$+2\text{cov}(\sigma_\beta^2, \sigma_e^2)] - 2\rho(1-\rho)[\text{cov}(\sigma_\beta^2, \sigma_\alpha^2) + \text{cov}(\sigma_\alpha^2, \sigma_e^2)]\Big\}, \tag{5.67}$$

$$\text{var}(\hat{\rho}) = \frac{1}{n(\sigma_\alpha^2 + \sigma_e^2)^2}\left[(1-\rho)^2\text{var}(\sigma_\alpha^2) + \rho^2\text{var}(\sigma_e^2) - 2\rho(1-\rho)\text{cov}(\sigma_\alpha^2, \sigma_e^2)\right], \tag{5.68}$$

and

$$\text{var}(\hat{\chi}_a) = \frac{1}{n(\sigma_\alpha^2 + \sigma_\beta^2 + \sigma_e^2)^2}\Big\{(1-\chi_a)^2[\text{var}(\sigma_\alpha^2) + \text{var}(\sigma_e^2) + 2\text{cov}(\sigma_\alpha^2, \sigma_e^2)]$$

$$+\chi_a^2\text{var}(\sigma_\beta^2) - 2\chi_a(1-\chi_a)[\text{cov}(\sigma_\beta^2, \sigma_\alpha^2) + \text{cov}(\sigma_\beta^2, \sigma_e^2)]\Big\}. \tag{5.69}$$

Again, we use the respective transformations for statistical inference for continuous data.

5.7.1   Other Estimation and Statistical Inference Approaches

Estimations of agreement indices shown in this chapter are based on the GEE methodology. We have discussed approaches by Barnhart, Song, and Haber (2005) throughout this chapter. Carrasco and Jover (2003) proposed another important method using the maximum likelihood (ML) or restricted maximum likelihood (RML) method based on a mixed effect model. In Section 2.11.3 we presented this model for $k = 2$, which is also applicable for $k > 2$. For normally distributed data, the estimates from Lin, Hedayat, and Wu (2007) have been shown to be very close to the estimates obtained in Carrasco and Jover (2003). However, the unified approach that we proposed has two advantages: First, since we use the GEE methodology to estimate all variance components and means, our method can handle not only normally distributed data, but also data from the exponential family including the binomial and multinomial distributions. Second, our approach is expected to be robust against a moderate deviation from normally distributed data.

King and Chinchilli (2001a) proposed one important method of estimations and statistical inferences for the generalized CCC for continuous and categorical data based on the U-statistic. The case $k = 2$ was presented in Section 2.11.1. We will present the case $k > 2$ below.

Let $q = 1, 2, \ldots, k - 1$ and $r = 2, \ldots, k$ index the pairwise combinations of the k raters. Then the generalized CCC can be expressed as

$$\bar{\rho}_g = \frac{\sum_{\substack{qr \\ q<r}} E_{F_{X_q} F_{X_r}} \left[g(X_q - X_r) - g(X_q + X_r) \right] - \sum_{\substack{qr \\ q<r}} E_{F_{X_q} X_r} \left[g(X_q - X_r) - g(X_q + X_r) \right]}{\sum_{\substack{qr \\ q<r}} E_{F_{X_q} F_{X_r}} \left[g(X_q - X_r) - g(X_q + X_r) \right] + \frac{1}{2} \sum_{\substack{qr \\ q<r}} E_{F_{X_q} X_r} \left[g(2X_q) + g(2X_r) \right]},$$

where $g(\cdot)$ is a distance function, including robust distance functions, defined in King and Chinchilli (2001a, 2001b), and F_{X_q} and F_{X_r} are the cumulative density functions (CDFs) of X_q and X_r. When $g(x) = x^2$, $\bar{\rho}_g$ becomes the CCC defined by Lin (1989).

Let $U_{qr} = (U_{1qr}, U_{2qr}, U_{3qr})'$ be the U-statistic as defined in Section 2.11.1. To construct U-estimators of the CCC with independent multivariate samples $(X_{11}, X_{21}, \ldots, X_{k1}), \ldots, (X_{1n}, X_{2n}, \ldots, X_{kn})$ of size n, King and Chinchilli (2001a) substitute F_{X_q}, F_{X_r}, and $F_{X_q X_r}$ by their respective empirical CDFs. For a general function g, this results in the following estimator of $\bar{\rho}_g$:

$$\hat{\bar{\rho}}_g = \frac{(n - 1) \sum_{\substack{qr \\ q<r}} (U_{3qr} - U_{1qr})}{\sum_{\substack{qr \\ q<r}} U_{1qr} + n \sum_{\substack{qr \\ q<r}} U_{2qr} + (n - 1) \sum_{\substack{qr \\ q<r}} U_{3qr}}.$$

Since the sum of U-statistics is a U-statistic, the above notation can be simplified as

$$\hat{\bar{\rho}}_g = \frac{(n - 1)(U_{3s} - U_{1s})}{U_{1s} + n U_{2s} + (n - 1) U_{3s}},$$

where $U_{ls} = \sum_{\substack{qr \\ q<r}} U_{lqr}$, $l = 1, 2, 3$, the sum over all distinct pairs. The same approach as found in Section 2.11.1 can be applied by replacing U_1, U_2, and U_3 with U_{1s}, U_{2s}, and U_{3s}, respectively. Then $\hat{\bar{\rho}}_g$ has an asymptotically normal distribution with mean $\bar{\rho}_g$ and a variance that can be consistently estimated with

$$\text{var}(\hat{\bar{\rho}}_g) \doteq (\hat{\bar{\rho}}_g)^2 \left[\frac{\text{var}(H_s)}{H_s^2} - \frac{2\text{cov}(H_s, G_s)}{H_s G_s} + \frac{\text{var}(G_s)}{G_s^2} \right].$$

As before, the normal approximation of the extended robust estimators of the concordance correlation coefficient can be improved through the use of the Z-transformation.

5.7.2 Variances of CCC and Weighted Kappa for $k = 2$

For ordinal and binary data, when $k = 2$ and $m = 1$, the above GEE estimates of CCC reduce to kappa (Cohen 1960) and weighted kappa (Cohen 1968) with square distance function (Robieson 1999; King and Chinchilli 2001a, 2001b; Barnhart, Haber, and Song 2002). In addition, its variances reduce to the variances of kappa and weighted kappa (Wu 2005). The proof of such equivalence is shown below.

When $k = 2$ and $m = 1$ with t ordinal outcomes, Fleiss, Cohen, and Everitt (1969) introduced the asymptotic variance for the estimated weighted kappa, $\hat{k}_w$, as

$$\text{var}(\hat{k}_w) = \frac{1}{n(1 - k_w)^4} \left(\sum_{i=1}^{t} \sum_{j=1}^{t} A_{ij} - B \right), \qquad (5.70)$$

where

$$A_{ij} = \pi_{ij} [w_{ij}(1 - \Pi_c) - (\bar{w}_{i\cdot} + \bar{w}_{\cdot j})(1 - \Pi_o)]^2, \qquad (5.71)$$

$$\bar{w}_{i\cdot} = \sum_{j=1}^{t} w_{ij} \pi_{\cdot j}, \qquad (5.72)$$

$$\bar{w}_{\cdot j} = \sum_{i=1}^{t} w_{ij} \pi_{i\cdot}, \qquad (5.73)$$

and

$$B = (\Pi_o \Pi_c - 2\Pi_c + \Pi_o)^2. \qquad (5.74)$$

In order to compare the variance of the weighted kappa to that of the CCC, we let

$$\mu_{l0} = \sum_{i=1}^{t} i^l \pi_{i\cdot}, \quad l = 1, 2, 3, 4,$$

$$\mu_{0l} = \sum_{i=1}^{t} i \pi_{\cdot j}^l, \quad l = 1, 2, 3, 4,$$

$$\mu_{ll'} = \sum_{i=1}^{t} \sum_{j=1}^{t} i^l j^{l'} \pi_{ij}, \quad l = 1, 2, 3 \text{ and } l' = 1, 2, 3$$

Based on the above notation, we obtain

$$1 - \Pi_o = \frac{1}{(t-1)^2} (\mu_{20} - 2\mu_{11} + \mu_{02}) \tag{5.75}$$

and

$$1 - \Pi_c = \frac{1}{(t-1)^2} (\mu_{20} - 2\mu_{01}\mu_{10} + \mu_{02}). \tag{5.76}$$

For the square weighted function w_{ij}, where

$$w_{ij} = 1 - \frac{(i-j)^2}{(t-1)^2}, \quad i, j = 1, 2, \ldots, t,$$

we have

$$\bar{w}_{i\cdot} = \sum_{j=1}^{t} w_{ij} \pi_{\cdot j}$$

$$= 1 - \frac{1}{(t-1)^2} \sum_{j=1}^{t} (i-j)^2 \pi_{\cdot j}$$

$$= 1 - \frac{1}{(t-1)^2} (i^2 - 2i\mu_{01} + \mu_{02}), \tag{5.77}$$

and

$$\bar{w}_{\cdot j} = \sum_{i=1}^{t} w_{ij} \pi_{i\cdot}.$$

$$= 1 - \frac{1}{(t-1)^2} \sum_{i=1}^{t} (i-j)^2 \pi_{i\cdot}.$$

$$= 1 - \frac{1}{(t-1)^2} (j^2 - 2j\mu_{10} + \mu_{20}). \tag{5.78}$$

By substituting all those terms into (5.70), we obtain

$$\text{var}(\hat{k}_w) = \frac{1}{n(1-\rho_c)^4}\left(\sum_{i=1}^{t}\sum_{j=1}^{t}A_{ij} - B\right)$$

$$= \frac{A1 + A2 - A3 - B1}{n}, \tag{5.79}$$

where

$$A1 = \frac{(t-1)^4 + \mu_{40} + 6\mu_{22} + \mu_{04} - 4\mu_{31} - 4\mu_{13} - 2(t-1)^2(\mu_{20} + \mu_{02} - 2\mu_{11})}{(\mu_{20} + \mu_{02} - 2\mu_{11})^2}, \tag{5.80}$$

$$A2 = \frac{4(t-1)^4(\mu_{20} + \mu_{20} - 2\mu_{11})^2 + (\mu_{40} + \mu_{04} + 3(\mu_{20} + \mu_{02})^2)(\mu_{20} + \mu_{02} - 2\mu_{11})^2}{(\mu_{20} + \mu_{02} - 2\mu_{10}\mu_{01})^4}$$

$$+ \frac{(4\mu_{10}^2\mu_{02} + 4\mu_{01}^2\mu_{20} - 4\mu_{10}\mu_{03} - 4\mu_{01}\mu_{30} + 2\mu_{22} - 4\mu_{10}\mu_{21})(\mu_{20} + \mu_{02} - 2\mu_{11})^2}{(\mu_{20} + \mu_{02} - 2\mu_{10}\mu_{01})^4}$$

$$+ \frac{(-4\mu_{01}\mu_{12} + 8\mu_{10}\mu_{01}\mu_{11} - 8\mu_{10}\mu_{01}\mu_{02} - 8\mu_{10}\mu_{20}\mu_{01})(\mu_{20} + \mu_{02} - 2\mu_{11})^2}{(\mu_{20} + \mu_{02} - 2\mu_{10}\mu_{01})^4}$$

$$+ \frac{-8(t-1)^2(\mu_{20} + \mu_{02} - 2\mu_{11})^2(\mu_{20} + \mu_{02} - 2\mu_{10}\mu_{01})}{(\mu_{20} + \mu_{02} - 2\mu_{10}\mu_{01})^4}, \tag{5.81}$$

$$A3 = \frac{(\mu_{20} + \mu_{02} - 2\mu_{11})}{(\mu_{20} + \mu_{02} - 2\mu_{10}\mu_{01})^3}\left[2(t-1)^4 - 2(t-1)^2(\mu_{20} + \mu_{02} - 2\mu_{11})\right.$$

$$-2(t-1)^2(\mu_{20} + \mu_{02} - 2\mu_{10}\mu_{01}) + \mu_{40} + \mu_{04} - 2\mu_{10}\mu_{03} - 2\mu_{01}\mu_{30} - 2\mu_{31} - 2\mu_{13}$$

$$+ (\mu_{20} + \mu_{02})^2 + 2\mu_{22} - 2\mu_{10}\mu_{21} - 2\mu_{01}\mu_{12} + 4\mu_{01}\mu_{21}$$

$$\left.+4\mu_{10}\mu_{12} - 2\mu_{20}\mu_{11} - 2\mu_{02}\mu_{11}\right] \tag{5.82}$$

and

$$B1 = \frac{(t-1)^8}{(\mu_{20} + \mu_{02} - 2\mu_{10}\mu_{01})^4}$$

$$\times \left\{\frac{[(t-1)^2 - (\mu_{20} + \mu_{02} - 2\mu_{11})][(t-1)^2 - (\mu_{20} + \mu_{02} - 2\mu_{10}\mu_{01})]}{(t-1)^4}\right.$$

$$\left.- \frac{2[(t-1)^2 - (\mu_{20} + \mu_{02} - 2\mu_{10}\mu_{01})]}{(t-1)^2} + \frac{[(t-1)^2 - (\mu_{20} + \mu_{02} - 2\mu_{11})]}{(t-1)^2}\right\}^2. \tag{5.83}$$

Using the system of equations as specified in (5.34) when $m = 1$, the variance of CCC can be expressed as

$$\text{var}(\hat{\rho}_c) = \frac{4}{n(\mu_{20} + \mu_{02} - 2\mu_{10}\mu_{01})^4}(A + B + C + D - E - F), \tag{5.84}$$

where

$$A = (\mu_{01}^2\mu_{20} + 2\mu_{10}\mu_{01}\mu_{11} + \mu_{10}^2\mu_{02} - 4\mu_{10}^2\mu_{01}^2)(\mu_{20} + \mu_{02} - 2\mu_{11})^2, \quad (5.85)$$

$$B = (\mu_{40} + 2\mu_{22} + \mu_{04} - (\mu_{20} + \mu_{02})^2)(\mu_{11} - \mu_{10}\mu_{01})^2, \quad (5.86)$$

$$C = (\mu_{22} - \mu_{11}^2)(\mu_{20} + \mu_{02} - 2\mu_{10}\mu_{01})^2, \quad (5.87)$$

$$D = 2(\mu_{11} - \mu_{10}\mu_{01})\left[\mu_{01}(\mu_{30} - \mu_{10}\mu_{20}) + \mu_{10}(\mu_{03} - \mu_{01}\mu_{02})\right.$$

$$\left. + \mu_{01}(\mu_{12} - \mu_{10}\mu_{02}) + \mu_{10}(\mu_{21} - \mu_{01}\mu_{20})\right](\mu_{20} + \mu_{02} - 2\mu_{11}), \quad (5.88)$$

$$E = 2(\mu_{20} + \mu_{02} - 2\mu_{11})(\mu_{20} + \mu_{02} - 2\mu_{10}\mu_{01})$$

$$\times(\mu_{01}\mu_{21} - 2\mu_{10}\mu_{01}\mu_{11} + \mu_{10}\mu_{12}), \quad (5.89)$$

and

$$F = 2(\mu_{31} + \mu_{13} - \mu_{11}\mu_{20} - \mu_{11}\mu_{02})(\mu_{11} - \mu_{10}\mu_{01})(\mu_{20} + \mu_{02} - 2\mu_{10}\mu_{01}). \quad (5.90)$$

Further simplifications yield

$$\mathrm{var}(\hat{\rho}_c) = \mathrm{var}(\hat{k}_w). \quad (5.91)$$

We have shown that when $k = 2$ and $m = 1$ and when the data are ordinal, the CCC without the Z-transformation is exactly the same as the weighted kappa with the squared weight function, both in estimation and statistical inference. For binary data, the weighted kappa reduces to kappa. Therefore, our approach can naturally extend the kappa and weighted kappa for $k > 2$ and $m \geq 1$. In addition, our approach provides precision and accuracy coefficients for categorical data.

5.8 Summary of Simulation Results

In order to evaluate the performance of the GEE methodology for estimation and inference of the proposed indices and to compare the proposed indices against other existing methods, simulation studies were designed and conducted for different types of data: binary, ordinal, and normal. For each of the three types of data, we considered three cases: $k = 2$ and $m = 1$, $k = 4$ and $m = 1$, and $k = 2$ and $m = 3$. For each case, we generated 1000 random samples of size 20 each. For binary and ordinal data, we considered two situations: inferences obtained through transformations (Z-transformations for CCC and precision indices, logit transformation for accuracy indices) and inferences obtained without transformations. For normal data, we considered only inferences obtained through transformations. In addition to the above transformation, we considered logit transformation for CP and log transformation for TDI for normal data.

For binary data, our estimates are very close to their corresponding theoretical values, and the means of the estimated standard deviations are very close to the corresponding standard deviations of the estimates. Therefore, our estimates are sufficiently good for binary data with or without transformation. When $m = 1$, we also compared our CCC estimates to that obtained from the method by Carrasco and Jover (2003). Our standard error estimates are superior to the estimates by Carrasco and Jover (2003) regardless of whether a transformation is used. The estimates obtained with transformation were comparable to the estimates obtained without transformation. Therefore, we suggest that for binary data, the use of a transformation is acceptable but not necessary.

For ordinal data, the means of the estimates are very close to the theoretical values, and the means of the estimated standard errors are very close to the corresponding standard deviations of the estimates. Similar to binary data, we also calculated the CCC estimates from the method by Carrasco and Jover (2003) for the cases with $m = 1$. The estimates from the two methods are very close to each other regardless of whether transformation is used. Therefore, we conclude that for ordinal data, the use of a transformation is acceptable but not necessary. Surprisingly, the method of Carrasco and Jover (2003) performs as well as ours for ordinal data, even though their model assumes normality.

For normal data, our estimates resemble the respective theoretical values very well. The means of the estimated standard error are very close to the corresponding standard deviations of the estimates. For the cases with $m = 1$, our CCCs are very close to that obtained from the method of Carrasco and Jover (2003). For the detailed simulation results, see Lin, Hedayat, and Wu (2007). Based on the simulation results, we conclude that our method works well for binary data, ordinal data, and normal data, both in estimates as well as in corresponding statistical inferences.

5.9 Examples

5.9.1 Example 1: Methods Comparison

Dispirin crosslinked hemoglobin (DCLHb) is a solution containing oxygen-carrying hemoglobin. The solution was created as a blood substitute to treat trauma patients and to replace blood loss during surgery. Measurements of DCLHb in patient's serum after infusion are routinely performed using a Sigma method. A method of measuring hemoglobin called the HemoCue photometer was modified to reproduce the Sigma instrument DCLHb results. To validate this modified method, serum samples from 299 patients over the analytical range of 50–2,000 mg/dL were collected. DCLHb values of each sample were measured simultaneously with the HemoCue and Sigma methods, andeach sample was measured twice by each of

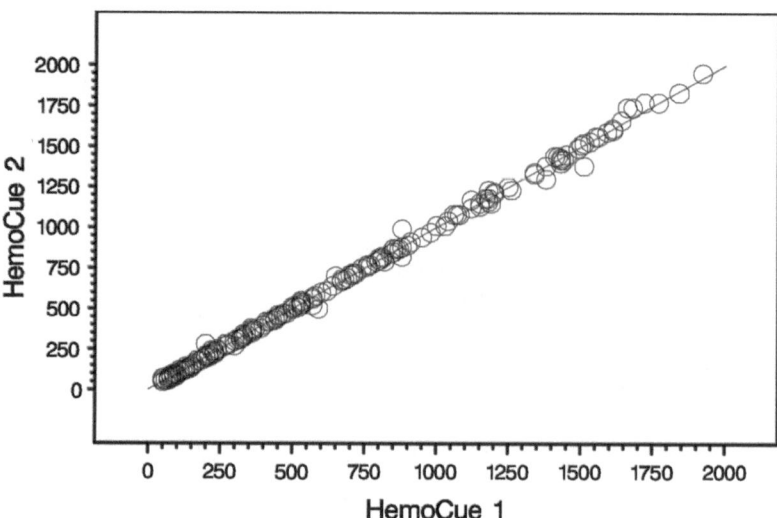

Fig. 5.1 HemoCue method measurement 1 vs. measurement 2

the two methods. This example has been given by Lin, Hedayat, Sinha, and Yang (2002) and Lin (2003), where the averages of the replicated readings were used, and is presented in Example 2.8.1.

Figures 5.1–5.3 plot the data for this example for the HemoCue method, measurement 1 vs. measurement 2; Sigma method, measurement 1 vs. measurement 2; and the average of the HemoCue method vs. the average of the Sigma method. The plots indicate that the errors are rather constant across the data range. Therefore, no log transformation was applied to the data.

In terms of TDI and CP indices, the least acceptable agreement is defined as having at least 90% of pair observations over the entire range within 75 mg/dL of each other if the observations are from the same method, and within 150 mg/dL of each other if the observations are from different methods based on the average of each method. In terms of CCC indices, the least acceptable agreement is defined as a within-sample total deviation of not more than 7.5% of the total deviation if observations are from the same method, and a within-sample total deviation of not more than 15% of the total deviation if observations are from different methods. These translate into a least-acceptable CCC_{intra} of $0.9943 = (1 - 0.075^2)$, and a least-acceptable CCC_{inter} of $0.9775 = (1 - 0.15^2)$.

The agreement statistics and their corresponding one-sided 95% lower or upper confidence limits are presented in Table 5.1. The CCC_{intra} estimate is 0.9986, which means for the observations from the same method that the within-sample deviation is about $3.7\% = \sqrt{1 - 0.9986}$ of the total deviation. The 95% lower confidence limit for CCC_{intra} is 0.9983, which is greater than 0.9943. The CCC_{inter} estimate is 0.9866, which means for the average observations from different methods, the within-sample total deviation is about $11.6\% = \sqrt{1 - 0.9866}$ of the

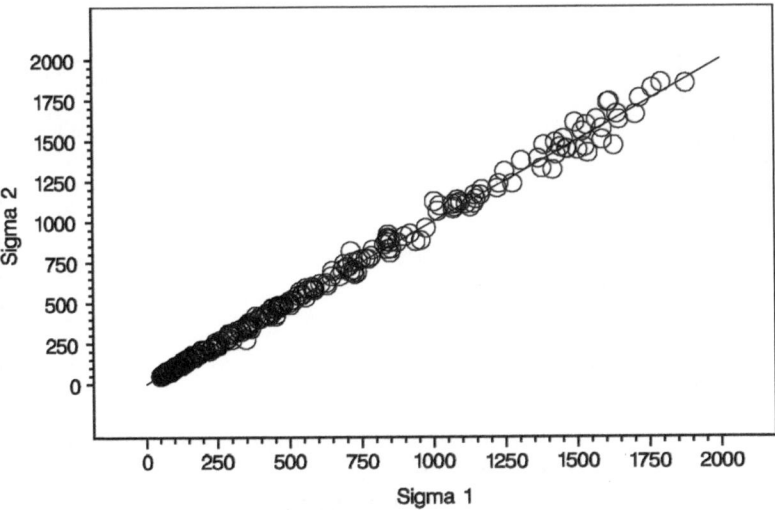

Fig. 5.2 Sigma method measurement 1 vs. measurement 2

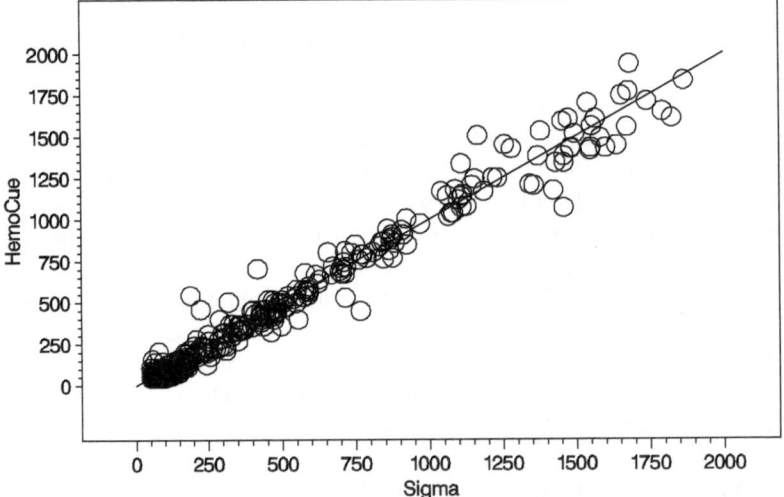

Fig. 5.3 HemoCue method's average measurement vs. Sigma method's average measurement

total deviation. The 95% lower confidence limit for CCC_{inter} is 0.9825, which is greater than 0.9775. The $precision_{intra}$ estimate is 0.9986, with a one-sided lower confidence limit 0.9983. The $precision_{inter}$ estimate is 0.9866 with a one-sided lower confidence limit 0.9825, and the $accuracy_{inter}$ estimate is 1.0000 with one-sided lower confidence limit 0.9987. The CCC_{total} estimate is 0.9859, which means for individual observations from different methods, the within-sample total deviation is

Table 5.1 Agreement statistics and their confidence limits for Example 5.9.1

Type	Statistics	CCC	Precision coefficient	Accuracy coefficient	$TDI_{0.9}$	CP_{TDI_a} [a]	RBS[b]
Intra	Estimate	0.9986	0.9986	.	41.1	0.9973	.
	95% Conf. limit	0.9983	0.9983	.	46.2	0.9949	.
	Allowance	0.9943	0.9943	.	75	0.9000	.
Inter	Estimate	0.9866	0.9866	1	127.3	0.9474	0
	95% Conf. limit	0.9825	0.9825	0.9987	145.9	0.9228	.
	Allowance	0.9775	.	.	150	0.9000	.
Total	Estimate	0.9859	0.9860	1	130.5	0.9412	0
	95% Conf. limit	0.9818	0.9818	0.9987	148.9	0.9160	.
	Allowance	0.9775	.	.	150	0.9000	.

For $k = 2$, $n = 299$, and $m = 2$.
[a]This is the CP given the TDI allowances of 75 mg/dL or 150 mg/dL.
[b]The relative bias squared (RBS) must be less than 1 or 8 for the CP criterion of 0.9 or 0.8, respectively, in order for the approximated TDI and CP to be valid.

about 11.87% of the total deviation. The 95% lower confidence limit for CCC_{total} is 0.9818. The precision$_{total}$ estimate is 0.9860 with a one-sided lower confidence limit 0.9818, and the accuracy$_{total}$ estimate is 1.0000 with one-sided lower confidence limit 0.9987.

The $TDI_{intra(0.9)}$ estimate is 41.1 mg/dL, which means that 90% of the readings are within 41.1 mg/dL of their replicate readings from the same method. The one-sided upper confidence limit for $TDI_{intra(0.9)}$ is 46.2 mg/dL, which is less than 75 mg/dL. The $TDI_{inter(0.9)}$ estimate is 127.3 mg/dL, which means that based on the average readings, 90% of the HemoCue readings are within 127.3 mg/dL of the Sigma readings. The one-sided upper confidence limit for $TDI_{inter(0.9)}$ is 145.9 mg/dL, which is slightly less than 150 mg/dL. The $TDI_{total(0.9)}$ estimate is 130.5 mg/dL, with the one-sided upper confidence limit 148.9 mg/dL, which is slightly less than 150 mg/dL as well.

Finally, the $CP_{intra(75)}$ estimate is 0.9973, which means that 99.7% of HemoCue observations are within 75 mg/dL of their duplicate values from the same method. The one-sided lower confidence limit for $CP_{intra(75)}$ is 0.9949, which is larger than 0.9. The $CP_{inter(150)}$ estimate is 0.9474, which means that 94.7% of HemoCue readings are within 150 mg/dL of the Sigma readings based on the average of each method. The one-sided lower confidence limit for $CP_{inter(150)}$ is 0.9228, which is larger than 0.9. The $CP_{total(150)}$ estimate is 0.9412, which means that 94% of HemoCue observations are within 150 mg/dL of Sigma observations based on individual readings. The one-sided lower confidence limit for $CP_{total(150)}$ is 0.9160.

The agreement between the HemoCue method and the Sigma method is acceptable with excellent accuracy and adequate precision and with accuracy slightly better than precision.

Table 5.2 Lab 1 frequency table of first reading (row) vs. second reading (column)

	Negative	Positive	Highly positive
Negative	6	1	0
Positive	0	49	0
Highly positive	0	0	8

Table 5.3 Lab 2 frequency table of first reading (row) vs. second reading (column)

	Negative	Positive	Highly positive
Negative	2	0	0
Positive	0	22	2
Highly positive	0	5	33

Table 5.4 Lab 1 first reading (row) vs. lab 2 first reading (column)

	Negative	Positive	Highly positive
Negative	2	5	0
Positive	0	19	30
Highly positive	0	0	8

Table 5.5 Lab 1 second reading (row) vs. lab 2 second reading (column)

	Negative	Positive	Highly positive
Negative	2	4	0
Positive	0	23	27
Highly positive	0	0	8

5.9.2   Example 2: Assay Validation

This example can be seen in Lin, Hedayat, and Wu (2007). In this example, we consider the hemagglutinin inhibition (HAI) assay for antibody to influenza A (H3N2) in rabbit serum samples from two different labs. Serum samples from 64 rabbits were measured twice by each method. Antibody level was classified as negative, positive, or highly positive (too numerous to count).

Tables 5.2–5.5 present the frequency tables for within-lab and between-lab readings. Tables 5.2 and 5.3 present the frequency tables of the first reading vs. the second reading from each lab. Table 5.4 presents the frequency table of the first reading from one lab vs. the first reading from the other lab. Table 5.5 presents the frequency table of the second reading from one lab vs. the second reading from the other lab. Those tables suggest that the within-lab agreement is good but the between lab agreement is not, and lab 2 tends to report higher ratings than lab 1.

This is an imprecise assay with ordinal responses, and therefore we allow for less-demanding agreement criteria. In terms of CCC indices, agreement was defined as a within-sample total deviation of not more than 50% of the total deviation if observations are from the same method, and a within-sample total deviation of not more than 75% of the total deviation if observations are from different methods. This translates into a least-acceptable CCC_{intra} of $0.75 = (1 - 0.5^2)$, and a least acceptable CCC_{inter} of $0.4375 = (1 - 0.75^2)$.

The estimates of agreement statistics and their corresponding one-sided 95% lower confidence limits are presented in Table 5.6. The CCC_{intra} is estimated to

Table 5.6 Agreement statistics and their confidence limits for Example 5.9.2

Type	Statistics	CCC	Precision coefficient	Accuracy coefficient
Intra	Estimate	0.8836	0.8836	.
	95% Conf. limit	0.8109	0.8109	.
	Allowance	0.7500	0.7500	.
Inter	Estimate	0.3723	0.5679	0.6554
	95% Conf. limit	0.2448	0.4571	0.5383
	Allowance	0.4375	.	.
Total	Estimate	0.3578	0.5349	0.6688
	95% Conf. limit	0.2335	0.4216	0.5570
	Allowance	.	.	.

For $k = 2$, $n = 64$, and $m = 2$.

be 0.8836, which means that for observations from the same method, the within-sample deviation is about $34.1\% = \sqrt{1 - 0.8836}$ of the total deviation. The 95% lower confidence limit for CCC_{intra} is 0.8109, which is larger than 0.7500. The CCC_{inter} is estimated to be 0.3723, which means that for the average observations from different methods, the within-sample deviation is about 79.2% of the total deviation. The 95% lower confidence limit for CCC_{inter} is 0.2448, which is less than 0.4375. The precision$_{inter}$ is estimated to be 0.5679 with a one-sided lower confidence limit 0.4571, and the accuracy$_{inter}$ is estimated to be 0.6554 with a one-sided lower confidence limit 0.5383. The CCC_{total} is estimated to be 0.3578, which means that for individual observations from different methods, the within-sample deviation is about 80.1% of the total deviation. The 95% lower confidence limit for CCC_{total} is 0.2335. The precision$_{total}$ is estimated to be 0.5349 with a one-sided lower confidence limit 0.4216, and the accuracy$_{total}$ is estimated to be 0.6688 with a one-sided lower confidence limit 0.5570.

Overall, the agreement between the two labs' readings is not acceptable, while the within-lab agreement is much better than the interlab agreement. The agreement within each lab, if of interest, can be obtained by applying kappa or weighted kappa to each lab separately.

5.9.3 Example 3: Nasal Bone Image Assessment by Ultrasound Scan

This example was obtained from Professor Philip Schluter, head of research, School of Public Health and Psychosocial Studies at AUT University, New Zealand. The leader of the project is Dr. Andrew McLennan, Sydney Ultrasound for Women, Sydney, Australia, and Royal North Shore Hospital, Sydney, Australia. This example will be published separately by McLennan and colleagues.

Several recent studies have demonstrated that the nasal bone (NB) is sonographically "absent" in a large proportion of fetuses affected by Down syndrome at

Table 5.7 Examiner 1: reading 1 (row) versus reading 2 (column)

	Absent	Present
Absent	316	14
Present	23	47

Table 5.8 Examiner 2: reading 1 (row) versus reading 2 (column)

	Absent	Present
Absent	318	9
Present	19	54

Table 5.9 Examiner 3: reading 1 (row) versus reading 2 (column)

	Absent	Present
Absent	285	9
Present	19	72

Table 5.10 Reading 1: examiner 1 (row) versus examiner 2 (column)

	Absent	Present
Absent	300	30
Present	27	43

Table 5.11 Reading 1: examiner 1 (row) versus examiner 3 (column)

	Absent	Present
Absent	274	56
Present	20	50

Table 5.12 Reading 1: examiner 2 (row) versus examiner 3 (column)

	Absent	Present
Absent	277	50
Present	17	56

11–13 weeks gestation. The purpose of this study was to demonstrate that the nasal bone can be accurately assessed and used for population screening in an Australian obstetric population at 11–13 weeks gestation.

There were 20 operators (accredited and experienced in nuchal translucency imaging) who supplied 20 NB images. The images were assessed for the presence (1) and absence (0) of NB. Three examiners assessed each of the 400 images twice (the repeat assessment separated by at least 24 h), giving a total of 2,400 assessments. Tables 5.7–5.12 present the frequency tables among three examiners and their duplicate readings. It appears that within-examiner has slightly better agreement than between-examiner, as expected, and there is little difference among the marginal distributions of three examiners (good accuracy).

This is an imprecise assay with binary responses, and therefore we allow for less-demanding agreement criteria. In terms of CCC indices, agreement was defined as a within-sample total deviation of no more than 60% of the total deviation if observations are from the same method based on the average of duplicate readings, and a within-sample total deviation of no more than 70% of the total deviation if observations are from different methods. This translates into a least-acceptable CCC_{intra} of $0.64 = (1 - 0.6^2)$, and a least acceptable CCC_{inter} of $0.51 = (1 - 0.7^2)$.

Table 5.13 Agreement statistics and their confidence limits for Example 5.9.3

Type	Statistics	CCC	Precision coefficient	Accuracy coefficient
Intra	Estimate	0.7047	0.7047	.
	95% Conf. limit	0.6558	0.6558	.
	Allowance	0.6400	0.6400	.
Inter	Estimate	0.6369	0.6442	0.9886
	95% Conf. limit	0.5779	0.5867	0.9811
	Allowance	0.5100	.	.
Total	Estimate	0.5438	0.5491	0.9902
	95% Conf. limit	0.4852	0.4913	0.9839
	Allowance	.	.	.

For $K = 3$, $n = 400$, and $m = 2$.

Table 5.13 presents the estimates of agreement statistics and their corresponding one-sided 95% lower confidence limits. The CCC_{intra} is estimated to be 0.7047, which means that for observations from the same method, the within-sample deviation is about 54.3% $= \sqrt{1 - 0.7047}$ of the total deviation. The 95% lower confidence limit for CCC_{intra} is 0.6558, which is better than 0.64. The CCC_{inter} is estimated to be 0.6369, which means that for the average observations from different methods, the within-sample deviation is about 60.3% of the total deviation. The 95% lower confidence limit for CCC_{inter} is 0.5779, which is better than 0.51. The precision$_{inter}$ is estimated to be 0.6442 with a one-sided lower confidence limit 0.5867, and the accuracy$_{inter}$ is estimated to be 0.9886 with a one-sided lower confidence limit 0.9811. The CCC_{total} is estimated to be 0.5438, which means that for individual observations from different methods, the within-sample deviation is about 67.5% of the total deviation. The 95% lower confidence limit for CCC_{total} is 0.4852. The precision$_{total}$ is estimated to be 0.5491 with a one-sided lower confidence limit 0.4913, and the accuracy$_{total}$ is estimated to be 0.9902 with a one-sided lower confidence limit 0.9839. Overall, the agreements among three examiners readings and within examiners are marginally acceptable, with very good accuracy, and most disagreements are from imprecision rather than inaccuracy.

5.9.4   Example 4: Accuracy and Precision of an Automatic Blood Pressure Meter

This example is obtained from Table 1 of Bland and Altman (1999), where a set of systolic blood pressure data from a study in which simultaneous measurements were made by each of two experienced observers (denoted by J and R) using a sphygmomanometer (gold standard) and by a semiautomatic blood pressure monitor (denoted by S). Three sets of readings were made in quick succession for each method (J_1–J_3, R_1–R_3, S_1–S_3). The purpose of the study was to evaluate whether the semiautomatic blood pressure monitor can replace the blood pressure apparatus

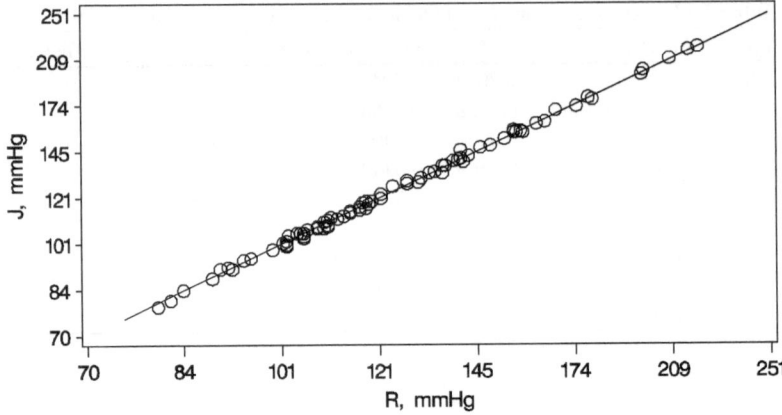

Fig. 5.4 Agreement between J and R based on their mean triplicate readings in log scale

routinely used in a typical medical center by an experienced nurse or a doctor. In their paper, the authors evaluated the agreement of only the first measurement by observer J and the machine (i.e., J_1 and S_1). This data set has $k = 3$ and $m = 3$. Because readings by J and R were almost identical based on mean of triplicate readings (see Fig. 5.4 in log scale), we analyzed this data set between J and S, or $k = 2$ and $m = 3$. Here we did not have any solid allowances (criteria) prespecified, and therefore, the allowances given are somewhat arbitrary.

First, we examine the distribution characteristic and the error structure to see whether we should assume proportional error when assessing agreement. We examine the S method because it is the most imprecise method. Figure 5.5 presents the plot of S_2 (circle) or S_3 (square) versus S_1 in the original scale. We can see that each marginal distribution is skewed to the right, and the error increases when the reading increases. Therefore, we proceed with assuming the proportional error assumption. In Figs. 5.6–5.9, we use the log scale with each tick mark multiplied by 1.2 from the previous tick mark.

Figure 5.6 presents the within-J agreement (precision) plot of J_2 (circle) or J_3 (square) versus J_1. Figure 5.7 presents the within-S agreement (precision) plot of S_2 (circle) or S_3 (square) versus S_1. We can see that the within-triplicate readings of J are more precise than those of S. Figure 5.8 presents the agreement plot of S_1 versus J_1 reflecting total agreement among individual readings. Figure 5.9 presents the agreement plot of S versus J based on the mean of triplicate readings reflecting interagreement. Figures 5.8 and 5.9 appear quite similar, indicating that the total agreement and interagreement are similar, while the within method had better agreement (precision), as expected. More importantly, the readings of the S method are neither accurate nor precise when compared to readings of the J method.

Table 5.14 presents the estimates of agreement statistics with 95% confidence limits. The 95% upper limit of within-method TDI%$_{0.9}$ is 15.5%, meaning that we are 95% confident that 90% of the within-method individual readings do not deviate

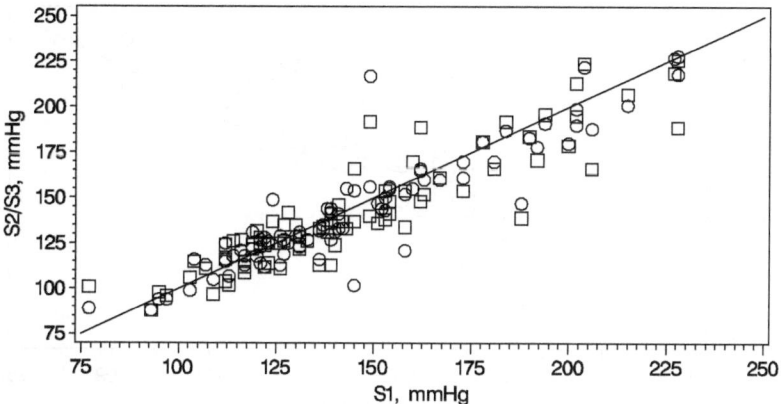

Fig. 5.5 Within-S meter agreement, S_2/S_3 vs. S_1 in original scale ($\circ = S_2, \square = S_3$)

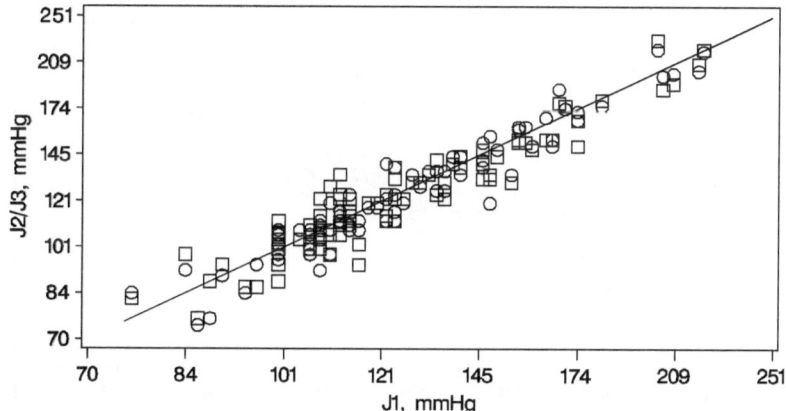

Fig. 5.6 Within-J meter agreement, J_2/J_3 vs. J_1 ($\circ = J_2, \square = J_3$) in log scale

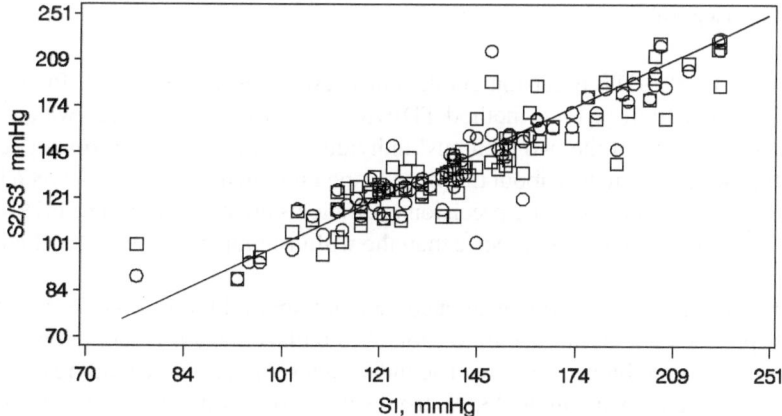

Fig. 5.7 Within-S meter agreement, S_2/S_3 vs. S_1 ($\circ = S_2, \square = S_3$) in log scale

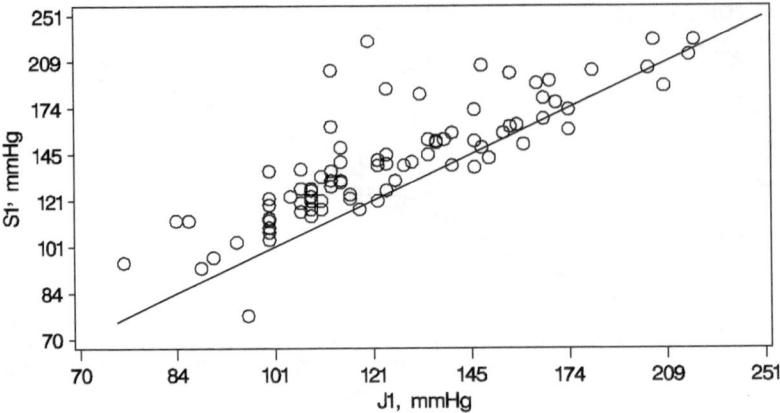

Fig. 5.8 Agreement between S_1 and J_1 (reflecting total agreement) in log scale

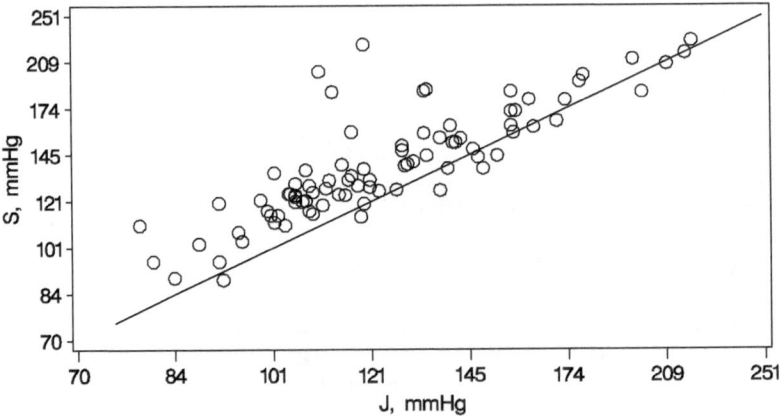

Fig. 5.9 Agreement between S and J based on their mean triplicate readings (reflecting intera-greement) in log scale

more than 15.5%. The precision coefficient is estimated to be 0.9383. In contrast, the 95% upper limit of total-method TDI%$_{0.9}$ is 43.5%, meaning that we are 95% confident that 90% of the two methods' individual triplicate readings do not deviate more than 43.5%, which is about the same as the inter method (41.3%) as seen in the figures. The total accuracy and precision coefficients are estimated to be 0.7974 and 0.8767, which are much less precise than the within-method precision coefficient of 0.9383.

We can conclude that this evaluated semiautomatic blood pressure monitor is neither precise enough nor accurate enough to replace the sphygmomanometer for measuring systolic blood pressure. The individual semiautomatic readings from the same patient can deviate up to 43.5% from sphygmomanometer readings measured by a nurse or doctor. Given the normal systolic pressure of 120 mmHg, the 43.5%

Table 5.14 Agreement statistics and their confidence limits for Example 5.9.4

Type	Statistics	CCC	Precision coefficient	Accuracy coefficient	$TDI_{0.9}$	CP_{TDI_a}	RBS[a]
Intra	Estimate	0.9383	0.9383	.	13.78	0.9798	.
	95% Conf. limit	0.9166	0.9166	.	15.46	0.9701	.
	Allowance	0.9000	0.9000	.	20	0.9000	.
Inter	Estimate	0.7253	0.8316	0.8721	33.05	0.8014	0.87
	95% Conf. limit	0.6044	0.7327	0.8132	41.34	0.7232	.
	Allowance	0.8	.	.	25	0.9000	.
Total	Estimate	0.6991	0.7974	0.8767	35.58	0.8438	0.69
	95% Conf. limit	0.5822	0.7015	0.8203	43.51	0.7831	.
	Allowance	0.7000	.	.	30	0.9000	.

For $K = 2$, $n = 85$, and $m = 3$.
[a]The relative bias squared (RBS) must be less than 1 or 8 for CP_a of 0.9 or 0.8, respectively, in order for the approximated TDI and CP to be valid.

deviation means that the result of the semiautomatic instrument can deviate by up to 52.2 mmHg with 95% confidence.

It is clear that the within-J precision (Fig. 5.6) is much better than the within-S precision (Fig. 5.7). To measure the agreement by each method, we can simply perform agreement assessment among the triplicate readings one method at a time with $k = 3$ and $m = 1$. However, this does not tell us whether within-J or within-R precision is significantly better than that of within-S. We will revisit this scenario in Chapter 6.

5.10   Discussion

We have proposed a series of indices for assessing agreement, precision, and accuracy for multiple raters each with multiple readings. Those indices can be used to assess intrarater, interrater, and total-rater agreement for both continuous and categorical data. These indices are summarized in Table 5.15. All those indices are expressed as functions of variance components through a two-way mixed model, and GEE methodology combined with the delta method is used to estimate all indices and perform their related statistical inferences. For sample size and power calculations of an agreement index, the reader is referred to the general procedures outlined in Section 4.1.

5.10.1   Relative or Scaled Indices

Each of the approaches in the previous chapters for assessing agreement becomes one of the special cases of our approach. For continuous data, when m approaches

Table 5.15 Summary of agreement indices based on functions of variance components

Statistics	Intra	Inter	Total	$m = 1$
CCC	$\dfrac{\sigma_\alpha^2 + \sigma_\gamma^2}{\sigma_\alpha^2 + \sigma_\gamma^2 + \sigma_e^2}$	$\dfrac{\sigma_\alpha^2}{\sigma_\alpha^2 + \sigma_\gamma^2 + \dfrac{\sigma_e^2}{m} + \sigma_\beta^2}$	$\dfrac{\sigma_\alpha^2}{\sigma_\alpha^2 + \sigma_\gamma^2 + \sigma_e^2 + \sigma_\beta^2}$	$\dfrac{\sigma_\alpha^2}{\sigma_\alpha^2 + \sigma_\beta^2 + \sigma_e^2}$
Precision	$\dfrac{\sigma_\alpha^2 + \sigma_\gamma^2}{\sigma_\alpha^2 + \sigma_\gamma^2 + \sigma_e^2}$	$\dfrac{\sigma_\alpha^2}{\sigma_\alpha^2 + \sigma_\gamma^2 + \dfrac{\sigma_e^2}{m}}$	$\dfrac{\sigma_\alpha^2}{\sigma_\alpha^2 + \sigma_\gamma^2 + \sigma_e^2}$	$\dfrac{\sigma_\alpha^2}{\sigma_\alpha^2 + \sigma_e^2}$
Accuracy	NA	$\dfrac{\sigma_\alpha^2 + \sigma_\gamma^2 + \dfrac{\sigma_e^2}{m}}{\sigma_\alpha^2 + \sigma_\gamma^2 + \dfrac{\sigma_e^2}{m} + \sigma_\beta^2}$	$\dfrac{\sigma_\alpha^2 + \sigma_\gamma^2 + \sigma_e^2}{\sigma_\alpha^2 + \sigma_\gamma^2 + \sigma_e^2 + \sigma_\beta^2}$	$\dfrac{\sigma_\alpha^2 + \sigma_e^2}{\sigma_\alpha^2 + \sigma_\beta^2 + \sigma_e^2}$
MSD	$2\sigma_e^2$	$2\sigma_\beta^2 + 2\sigma_\gamma^2 + 2\dfrac{\sigma_e^2}{m}$	$2\sigma_\beta^2 + 2\sigma_\gamma^2 + 2\sigma_e^2$	$2\sigma_e^2 + 2\sigma_\beta^2$
TDI_{π_0} [a]	$Q\sqrt{MSD_{Intra}}$	$Q\sqrt{MSD_{inter}}$	$Q\sqrt{MSD_{total}}$	$Q\sqrt{MSD}$
CP_{δ_0} [b]	$\chi^2\left(\dfrac{\delta^2}{MSD_{Intra}}, 1\right)$	$\chi^2\left(\dfrac{\delta^2}{MSD_{inter}}, 1\right)$	$\chi^2\left(\dfrac{\delta^2}{MSD_{total}}, 1\right)$	$\chi^2\left(\dfrac{\delta^2}{MSD}, 1\right)$

[a] $Q = \Phi^{-1}(1 - \frac{1-\pi}{2})$ is the inverse cumulative normal distribution.

[b] $\chi^2(\frac{\delta^2}{MSD}, 1)$ is a central chi-square distribution with one degree of freedom.

infinity, the proposed CCC_{inter} reduces to that proposed by Barnhart, Song, and Haber (2005). When $m = 1$, the proposed CCC reduces to the CCC proposed by Carrasco and Jover (2003), which is the same as the OCCC proposed by Lin (1989), King and Chinchilli (2001a), and Barnhart, Haber, and Song (2002). Barnhart, Haber, and Song (2002) pointed out that OCCC is actually a weighted average of pairwise CCC values. When $k = 2$ and $m = 1$, the proposed CCC reduces to the original CCC proposed by Lin (1989). For categorical data, when $k = 2$ and $m = 1$, the proposed CCC reduces to the kappa for binary data and weighted kappa with squared weight for ordinal data, in both estimates and statistical inferences.

In addition, we decomposed the CCC into precision and accuracy components for a deeper understanding of the sources of the disagreement. The concept of accuracy and precision can also be applied to categorical data. For continuous data, the relative or scaled indices are heavily dependent on the total variability (total data range). Therefore, these indices are not comparable if the ranges of the data are not comparable. The same is true for categorical data when we have data that are heavily clustered into a single cell, for example, when evaluating agreement based on low prevalence rate.

5.10.2   Absolute or Unscaled Indices

We also have proposed absolute indices, MSD, TDI, and CP, which are independent of the total data range. These absolute indices are easily comprehensible. However, these absolute indices are valid only when the relative bias squared is small enough (Lin 2000, 2003; Lin, Hedayat, Sinha, and Yang 2002) and the normality is assumed.

5.10.3   Covariate Adjustment

We refer the reader to Section 2.12.7, most of which is applicable to this chapter as well. Subject-based covariates can conveniently be adjusted using the model

$$Y_{ijl} - X_i\eta = \mu + \alpha_i + \beta_j + \gamma_{ij} + e_{ijl}, \tag{5.92}$$

where $X_i = (x_{i1}, x_{i2}, \ldots, x_{ip})'$ represents p covariates for the subject or sample i without the intercept term, and $\eta = (\eta_1, \eta_2, \ldots, \eta_p)'$ represents slopes of the p covariates.

A simple and reasonable approach is to perform linear regression for each rater j and replicate l, then use the intercept estimate plus the residual of each rater j and replicate l as the adjusted dependent variable, $Z_{ijl} = Y_{ijl} - X_i\eta$. We can then proceed to perform the estimations and statistical inferences of the agreement indices based on model (5.1) using the adjusted dependent variables. A formal and more efficient way is to include the $X_i\eta$ term in model (5.1), and solve for the GEE estimates and their variance–covariance matrix iteratively.

5.10.4   Future Research Topics and Related Publications

There are two aspects of this unified approach that can be extended and developed. First, for categorical and nonnormal continuous data, we may include the link functions, such as log and logit, in the GEE methodology. We expect the approach with a link function would become more robust to different types of data. Second, current variance component functions are based on balanced data. Therefore, we would have to delete samples or subjects with missing data. Approaches that can handle missing data should be an interesting area of research.

There are relatively few references available related to this chapter in medical and pharmaceutical publications other than those mentioned in the first paragraph of this chapter. See Barnhart, Haber, and Lin (2007) for an overview on assessing agreement with continuous measurements. Chen and Barnhart (2008) compared ICC and CCC for assessing agreement for data without and with replications. Haber and Barnhart (2008) proposed a general approach to evaluating agreement between two raters or methods of measurement from quantitative data with replicated measurements. There are many references in the social and psychological sciences for ICC-type indices. Some of these indices are closely related to the relative indices shown in this chapter. These are well represented in Brennan (2001).

Chapter 6
A Comparative Model for Continuous and Categorical Data

In Chapter 5, we provided statistical tools for assessing the intra-, inter-, and total-rater agreement among all raters. In this chapter, we provide statistical tools for comparing total-rater agreement to intrarater precision, and intrarater precision among selected raters.

When multiple raters are available with replicates, we are often interested to know whether raters can be used interchangeably if they do not deviate too much more than they deviate among their replicates without any clinical or practical alteration. Here, we need to assume that the variation among replicates (intrarater variability) is acceptable. FDA's guideline (2001) (http://www.fda.gov/downloads/Drugs/GuidanceComplianceRegulatoryInformation/Guidances/ucm070244.pdf) introduced a method for evaluating individual agreement between a test drug and a reference drug in the context of individual bioequivalence. Barnhart, Kosinski, and Haber (2007) extended FDA's approach in the case of multiple raters. They proposed the individual equivalence coefficient (IEC) and the coefficient of individual agreement (CIA) to compare the intrarater precision relative to the total-rater agreement of one or multiple references, or when no reference is available. They used the method of moments for estimation and nonparametric bootstrapping for statistical inference. Our approaches allow users to explore total-rater agreement relative to intrarater precision, whereby users can select raters of interest to be evaluated. We also allow users to compare intrarater precision among selected raters. We present such a general comparison model and propose the total–intra ratio (TIR) as well as the intra–intra ratio (IIR) to evaluate the comparative agreement when there exists a reference and when there does not.

The TIR is a noninferiority assessment such that the differences of individual readings from different raters can not be inferior by a certain margin to the differences of the replicated readings within raters. For a TIR example, to assess the individual bioequivalence, the agreement of test and reference compound is assessed relative to the agreement of within-reference compound. Although the TIR is equivalent to IEC and CIA, we have proposed on alternative statistical inference approach using the GEE methodology that works for both continuous and categorical data.

L. Lin et al., *Statistical Tools for Measuring Agreement*,
DOI 10.1007/978-1-4614-0562-7_6, © Springer Science+Business Media, LLC 2012

The IIR is a classical assessment for which the precision of selected assays/raters can be better than, equal to, or worse than that of other assays/raters. For an IIR example, in the medical-device environment, we often want to know whether the within-device precision of a newly developed device is better than, equal to, or worse than that of the within-device precision for the old device.

GEE methodology is used for estimation and statistical inference. Our approach allows for selecting any subset of raters as test and reference raters. In addition, we present assorted examples to demonstrate the flexibility of our approach. More details about this chapter can be found in Lin, Hedayat, and Tang (2012).

6.1   General Model

Suppose each of n randomly selected subjects is measured by each of k fixed raters with m replications. The general model for comparative agreement is

$$y_{ijl} = \mu_{ij} + e_{ijl}. \tag{6.1}$$

Here y_{ijl} represents the lth reading by rater j for subject i, with $i = 1, 2, \ldots, n$, $j = 1, 2, \ldots, k$, and $l = 1, 2, \ldots, m$. The true reading of the jth rater on subject i is μ_{ij}, and it is considered random because subjects are considered random, although the mean of each rater, μ_j, is considered fixed. The residual random effect is e_{ijl}. We make the following assumptions:

$$\mathrm{E}(\mu_{ij}) = \mu_j,$$

$$\mathrm{var}(\mu_{ij}) = \lambda_j^2,$$

$$\mathrm{corr}(\mu_{ij}, \mu_{ij'}) = \rho_{jj'},$$

$$\mathrm{E}(e_{ijl}) = 0, \text{ and}$$

$$\mathrm{var}(e_{ijl}) = \varsigma_j^2.$$

Here, we assume that the e_{ijl} are uncorrelated, μ_{ij} and e_{ijl} are uncorrelated, and replicates within a rater are interchangeable.

We use model (6.1) as opposed to model (5.1) because we allow the flexibility of evaluating any subset of raters in this chapter. For continuous data, when the within-sample error is proportional to the observed data values, we apply a log transformation to the data. Based on model (6.1), we propose to use mean squared deviation (MSD) to assess comparative agreement.

6.2 MSD for Continuous and Categorical Data

6.2.1 Intrarater Precision

Under the assumption that replicates within a rater are interchangeable, we propose to use $\mathrm{MSD_{intra}}$ to evaluate the intrarater agreement, which is the mean squared deviation among replicated readings within raters. Intrarater agreement evaluates the precision within raters. For a chosen rater j, based on model (6.1), $\mathrm{MSD_{intra}}_j$ can be expressed as

$$
\begin{aligned}
\varepsilon^2_{\mathrm{intra}_j} &= E(y_{ijl} - y_{ijl'})^2 \\
&= (\mu_j - \mu_j)^2 + (\lambda_j^2 + \varsigma_j^2 + \lambda_j^2 + \varsigma_j^2) - 2\lambda_j^2 \\
&= 2\varsigma_j^2.
\end{aligned}
\tag{6.2}
$$

For two or more raters, $\mathrm{MSD_{intra}}$ is then defined as the average of $\mathrm{MSD_{intra}}_j$ for the selected s raters.

6.2.2 Total-Rater Agreement

For evaluating total-rater agreement, we proposed in Chapter 5 to use $\mathrm{MSD_{total}}$, which is based on any individual reading from each rater. For any two raters, say raters j and j', based on model (6.1), $\mathrm{MSD_{total}}_{jj'}$ can be expressed as

$$
\begin{aligned}
\varepsilon^2_{\mathrm{total}_{jj'}} &= E(y_{ijl} - y_{ij'l'})^2 \\
&= (\mu_j - \mu_{j'})^2 + (\lambda_j^2 + \varsigma_j^2 + \lambda_{j'}^2 + \varsigma_{j'}^2) - 2\rho_{jj'}\lambda_j\lambda_{j'}.
\end{aligned}
\tag{6.3}
$$

Overall, $\mathrm{MSD_{total}}$ for the selected s raters is then defined as the average across $s(s-1)/2$ pairs of $\mathrm{MSD_{total}}_{jj'}$.

6.2.3 Interrater Agreement

In Chapter 5, we evaluated the interrater agreement using $\mathrm{MSD_{inter}}$. In this chapter, however, $\mathrm{MSD_{inter}}$ need not be evaluated, because $\mathrm{MSD_{total}}$ and $\mathrm{MSD_{intra}}$ include all the information for evaluating comparative agreement. For any two raters, say rater j and j', based on the model, MSD$_{\text{inter}_{jj'}}$ can be expressed as

$$\varepsilon^2_{\text{inter}_{jj'}} = E(\bar{y}_{ij\cdot} - \bar{y}_{ij'\cdot})^2$$

$$= (\mu_j - \mu_{j'})^2 + \left(\lambda_j^2 + \frac{\varsigma_j^2}{m} + \lambda_{j'}^2 + \frac{\varsigma_{j'}^2}{m}\right) - 2\rho_{jj'}\lambda_j\lambda_{j'}. \quad (6.4)$$

The overall MSD$_{\text{inter}}$ for the selected s raters is the average across $s(s-1)/2$ pairs of MSD$_{\text{inter}_{jj'}}$. The relationship between MSD$_{\text{intra}}$, MSD$_{\text{inter}}$, and MSD$_{\text{total}}$ can be obtained upon simplification:

$$\frac{\varepsilon^2_{\text{total}}}{\varepsilon^2_{\text{intra}}} = \frac{\varepsilon^2_{\text{inter}}}{\varepsilon^2_{\text{intra}}} + \left(1 - \frac{1}{m}\right). \quad (6.5)$$

The above equation shows that as soon as MSD$_{\text{total}}$ and MSD$_{\text{intra}}$ are determined, then MSD$_{\text{inter}}$ is given as well. Therefore, for comparative agreement, further evaluation for MSD$_{\text{inter}}$ is not necessary.

For comparative agreement, the exact intrarater agreement indices TDI and CP, as shown in Chapter 5, are one-to-one functions of MSD$_{\text{intra}}$ for normally distributed data. The approximate total TDI and CP are one-to-one functions of MSD$_{\text{total}}$ for normally distributed data. Within the same experiment, the between-sample variances or data ranges are similar, and therefore further scaling relative to the between-sample variance such as CCC is not necessary for comparative agreement. Furthermore, such scaling would reduce the power of the experiment.

6.2.4 Categorical Data

Let any two replicates from the same rater or different raters, denoted by X and Y, represent the classification scores of a subject in one of t categories. Table 3.1 presents the agreement table for all possible probability outcomes, where π_{pq} represents the probability of $X = p$ and $Y = q$, $p, q = 1, 2, \ldots, t$.

Based on the agreement probabilities presented in Table 3.1, when X and Y represent any two replicates within rater j, MSD$_{\text{intra}_j}$ for categorical data becomes

$$\varepsilon^2_{\text{intra}_j} = \sum_p \sum_q (p-q)^2 \pi_{pq}, \quad p, q = 1, 2, \ldots, t. \quad (6.6)$$

Let Π_{0_j} be the weighted probability of agreement within rater j with the squared weight function defined in (3.3). Then $\Pi_{0_j} = \sum_p^t \sum_q^t w_{pq}\pi_{pq}$. Therefore, the relationship between MSD$_{\text{intra}_j}$ and the weighted probability of agreement becomes

$$\varepsilon^2_{\text{intra}_j} = (t-1)^2(1 - \Pi_{0_j}). \quad (6.7)$$

Note that weighted kappa is the chance corrected scaling of the weighted probability of agreement.

Similarly, when X and Y represent any replicate from raters j and j', respectively, $\mathrm{MSD}_{\mathrm{total}_{jj'}}$ for categorical data becomes

$$\varepsilon^2_{\mathrm{total}_{jj'}} = \sum_p \sum_q (p-q)^2 \pi_{pq} = (t-1)^2 (1 - \Pi_{0_{jj'}}), \quad p, q = 1, 2, \ldots, t,$$

(6.8)

where $\Pi_{0_{jj'}}$ is the weighted probability of agreement from any replicate of raters j and j'. For categorical data, kappa and weighted kappa are the most common indices for assessing agreement between two raters, each with a single reading. Because of the equivalence of CCC and weighted kappa, weighted kappa is also largely dependent on MSD. Therefore, within the same experiment, further scaling from MSD to weighted kappa is not necessary for evaluating the comparative agreement.

6.3   GEE Estimation

Let $\theta = (\mu, \varsigma^2, \lambda^2, \rho)'$ be the vector of parameters with $\mu = (\mu_1, \ldots, \mu_j, \ldots, \mu_k)'$, $\varsigma^2 = (\varsigma_1^2, \ldots, \varsigma_j^2, \ldots, \varsigma_k^2)'$, $\lambda^2 = (\lambda_1^2, \ldots, \lambda_j^2, \ldots, \lambda_k^2)'$, and $\rho = (\rho_{12}, \ldots, \rho_{jj'}, \ldots, \rho_{(k-1)k})'$. Estimates of θ along with its variance–covariance matrix can be obtained via GEE methodology. The following system of estimation equations is used to obtain the estimates of the parameters.

We estimate μ by the first set of estimating equations:

$$\sum_{i=1}^{n} F'_{i1} H_{i1}^{-1} (Q_{i1} - \mu) = 0,$$

(6.9)

where

$$Q_{i1} = \begin{pmatrix} (y_{i11} + y_{i12} + \cdots + y_{i1m})/m \\ \vdots \\ (y_{ij1} + y_{ij2} + \cdots + y_{ijm})/m \\ \vdots \\ (y_{ik1} + y_{ik2} + \cdots + y_{ikm})/m \end{pmatrix},$$

$$\mu = \begin{pmatrix} \mu_1 \\ \vdots \\ \mu_j \\ \vdots \\ \mu_k \end{pmatrix},$$

and $F_{i1} = \frac{\partial \mu}{\partial(\mu_1,\ldots,\mu_k)} = I_{k \times k}$. The working covariance matrix for Q_{i1} (Zeger and Liang 1986) is

$$H_{i1} = \text{diag}(\text{var}(Q_{i1})) = \begin{pmatrix} a_1^{(1)} & & & & \\ & \ddots & & \mathbf{0} & \\ & & a_j^{(1)} & & \\ & \mathbf{0} & & \ddots & \\ & & & & a_k^{(1)} \end{pmatrix},$$

where

$$a_j^{(1)} = \text{var}\left[\frac{1}{m}(y_{ij1} + y_{ij2} + \cdots + y_{ijm})\right]$$

$$= \lambda_j^2 + \frac{\varsigma_j^2}{m}. \tag{6.10}$$

Note that we assume normality for constructing all of the working covariance matrices. The estimator for μ is

$$\hat{\mu} = \left(\sum_{i=1}^{n} F_{i1}' H_{i1}^{-1} F_{i1}\right)^{-1} \left(\sum_{i=1}^{n} F_{i1}' H_{i1}^{-1} Q_{i1}\right). \tag{6.11}$$

We estimate ς^2 by the second set of estimating equations:

$$\sum_{i=1}^{n} F_{i2}' H_{i2}^{-1} (Q_{i2} - \varsigma^2) = 0, \tag{6.12}$$

where

$$Q_{i2} = \begin{pmatrix} \sum_{l=1}^{m}(y_{i1l} - \bar{y}_{i1.})^2/(m-1) \\ \vdots \\ \sum_{l=1}^{m}(y_{ijl} - \bar{y}_{ij.})^2/(m-1) \\ \vdots \\ \sum_{l=1}^{m}(y_{ikl} - \bar{y}_{ik.})^2/(m-1) \end{pmatrix},$$

$$\varsigma^2 = \begin{pmatrix} \varsigma_1^2 \\ \vdots \\ \varsigma_j^2 \\ \vdots \\ \varsigma_k^2 \end{pmatrix},$$

and $F_{i2} = \frac{\partial \varsigma^2}{\partial(\varsigma_1^2, \ldots, \varsigma_k^2)} = I_{k \times k}$. The working covariance matrix for Q_{i2} is

$$H_{i2} = \text{diag}(\text{var}(Q_{i2})) = \begin{pmatrix} a_1^{(2)} & & & & \\ & \ddots & & \mathbf{0} & \\ & & a_j^{(2)} & & \\ & \mathbf{0} & & \ddots & \\ & & & & a_k^{(2)} \end{pmatrix},$$

where

$$a_j^{(2)} = \text{var}\left[\frac{\sum_{l=1}^{m} (y_{ijl} - \bar{y}_{ij\cdot})^2}{m-1} \right]$$

$$= \frac{2\varsigma_j^2}{m-1}. \tag{6.13}$$

The estimator for ς^2 is

$$\hat{\varsigma}^2 = \left(\sum_{i=1}^{n} F_{i2}' H_{i2}^{-1} F_{i2} \right)^{-1} \left(\sum_{i=1}^{n} F_{i2}' H_{i2}^{-1} Q_{i2} \right). \tag{6.14}$$

We estimate λ^2 by the third set of estimating equations:

$$\sum_{i=1}^{n} F_{i3}' H_{i3}^{-1} (Q_{i3} - g(\mu, \varsigma^2, \lambda^2)) = 0, \tag{6.15}$$

where

$$Q_{i3} = \begin{pmatrix} (y_{i11} + y_{i12} + \cdots + y_{i1m})^2 / m^2 \\ \vdots \\ (y_{ij1} + y_{ij2} + \cdots + y_{ijm})^2 / m^2 \\ \vdots \\ (y_{ik1} + y_{ik2} + \cdots + y_{ikm})^2 / m^2 \end{pmatrix},$$

$$g(\mu, \varsigma^2, \lambda^2) = \begin{pmatrix} \mu_1^2 + \varsigma_1^2 / m + \lambda_1^2 \\ \vdots \\ \mu_j^2 + \varsigma_j^2 / m + \lambda_j^2 \\ \vdots \\ \mu_k^2 + \varsigma_k^2 / m + \lambda_k^2 \end{pmatrix},$$

and $F_{i3} = \frac{\partial g(\mu, \varsigma^2, \lambda^2)}{\partial(\lambda_1^2, \ldots, \lambda_k^2)}$. The working covariance matrix for Q_{i3} is

$$H_{i3} = \mathrm{diag}(\mathrm{var}(Q_{i3})) = \begin{pmatrix} a_1^{(3)} & & & & \\ & \ddots & & \mathbf{0} & \\ & & a_j^{(3)} & & \\ & \mathbf{0} & & \ddots & \\ & & & & a_k^{(3)} \end{pmatrix},$$

where

$$a_j^{(3)} = \mathrm{var}\left[\frac{1}{m^2}(y_{ij1} + y_{ij2} + \cdots + y_{ijm})^2\right]$$

$$= 2\left(\lambda_j^2 + \frac{\varsigma_j^2}{m}\right)^2 + 4\mu_j^2\left(\lambda_j^2 + \frac{\varsigma_j^2}{m}\right). \tag{6.16}$$

We obtain the estimate for μ and ς^2 from (6.11) and (6.14), namely, $\hat{\mu}$ and $\hat{\varsigma}^2$. We then solve for $\hat{\lambda}^2$ using the equation

$$\hat{\lambda}^2 = \left(\sum_{i=1}^{n} F_{i3}' H_{i3}^{-1} F_{i3}\right)^{-1}\left(\sum_{i=1}^{n} F_{i3}' H_{i3}^{-1} Q_{i3}\right) - \hat{\mu}^2 - \frac{\hat{\varsigma}^2}{m}, \tag{6.17}$$

where $\hat{\mu}^2$ represents the vector in which the element is the square of the corresponding element of the vector $\hat{\mu}$.

Finally, we estimate ρ by the fourth set of estimating equations using the cross products:

$$\sum_{i=1}^{n} F_{i4}' H_{i4}^{-1}(Q_{i4} - h(\mu, \lambda^2, \rho)) = 0, \tag{6.18}$$

where

$$Q_{i4} = \begin{pmatrix} \bar{y}_{i1\cdot} \bar{y}_{i2\cdot} \\ \vdots \\ \bar{y}_{i1\cdot} \bar{y}_{ik\cdot} \\ \bar{y}_{i2\cdot} \bar{y}_{i3\cdot} \\ \vdots \\ \bar{y}_{i2\cdot} \bar{y}_{ik\cdot} \\ \vdots \\ \bar{y}_{ij\cdot} \bar{y}_{ij'\cdot} \\ \vdots \\ \bar{y}_{i(k-1)\cdot} \bar{y}_{ik\cdot} \end{pmatrix}_{1 \times \frac{k(k-1)}{2}},$$

$$h(\boldsymbol{\mu}, \boldsymbol{\lambda}^2, \boldsymbol{\rho}) = \begin{pmatrix} \mu_1 \mu_2 + \rho_{12} \lambda_1 \lambda_2 \\ \vdots \\ \mu_j \mu_{j'} + \rho_{jj'} \lambda_j \lambda_{j'} \\ \vdots \\ \mu_{(k-1)} \mu_k + \rho_{(k-1)k} \lambda_{k-1} \lambda_k \end{pmatrix}_{1 \times \frac{k(k-1)}{2}},$$

and $F_{i4} = \frac{\partial h(\boldsymbol{\mu}, \boldsymbol{\lambda}^2, \boldsymbol{\rho})}{\partial(\rho_{12}, \ldots, \rho_{(k-1)k})}$. The working covariance matrix for Q_{i4} is

$$H_{i4} = \mathrm{diag}(\mathrm{var}(Q_{i4}))$$

$$= \begin{pmatrix} a_{12}^{(4)} & & & & & \\ & \ddots & & & \mathbf{0} & \\ & & a_{1k}^{(4)} & & & \\ & & & a_{23}^{(4)} & & \\ & & & & \ddots & \\ & & & & & a_{2k}^{(4)} & \\ & & & & & & \ddots \\ & \mathbf{0} & & & & & & a_{jj'}^{(4)} \\ & & & & & & & & \ddots \\ & & & & & & & & & a_{(k-1)k}^{(4)} \end{pmatrix}_{\frac{k(k-1)}{2} \times \frac{k(k-1)}{2}},$$

where

$$a_{jj'}^{(4)} = \mathrm{var}(\bar{y}_{ij\cdot} \bar{y}_{ij'\cdot})$$

$$= \left(\rho_{jj'} \lambda_j^2 \lambda_{j'}^2 \right)^2 + 2 \mu_j \mu_{j'} \rho_{jj'} \lambda_j^2 \lambda_{j'}^2$$

$$+ \left(\lambda_j^2 + \varsigma_j^2 \right) \left(\lambda_{j'}^2 + \varsigma_{j'}^2 \right) + \mu_j \left(\lambda_{j'}^2 + \frac{\varsigma_{j'}^2}{m} \right) + \mu_{j'} \left(\lambda_j^2 + \frac{\varsigma_j^2}{m} \right). \quad (6.19)$$

We obtain the estimate for $\boldsymbol{\lambda}^2$ from (6.17), namely $\hat{\boldsymbol{\lambda}}^2$. We then use $\hat{\boldsymbol{\mu}}$ and $\hat{\boldsymbol{\lambda}}^2$ in (6.18) for $\boldsymbol{\mu}$ and $\boldsymbol{\lambda}^2$ and solve for $\hat{\boldsymbol{\rho}}$. Upon simplification, the estimate for each element of $\hat{\boldsymbol{\rho}}$ is given by

$$\hat{\rho}_{jj'} = \left(\frac{\sum_{i=1}^n \bar{y}_{ij\cdot} \bar{y}_{ij'\cdot}}{n} - \hat{\mu}_j \hat{\mu}_{j'} \right) \Big/ \hat{\lambda}_j \hat{\lambda}_{j'}, \quad (6.20)$$

where $\hat{\mu}_j$ and $\hat{\mu}_{j'}$ are the jth and j'th elements of vector $\hat{\boldsymbol{\mu}}$, and $\hat{\lambda}_j$ and $\hat{\lambda}_{j'}$ are the square roots of the jth and j'th elements of vector $\hat{\boldsymbol{\lambda}}^2$.

When there is no covariate, no iteration is needed for the above estimation process. The variance–covariance matrix for the estimated parameters $\hat{\theta} = (\hat{\mu}, \hat{\varsigma}^2, \hat{\lambda}^2, \hat{\rho})'$ is given by

$$V(\hat{\theta}) = \frac{1}{n} D^{-1'} \Sigma D^{-1}, \qquad (6.21)$$

where

$$
D = \begin{pmatrix}
\sum_{i=1}^{n} F_{i1}' H_{i1}^{-1} F_{i1} & 0 & 0 & 0 \\
0 & \sum_{i=1}^{n} F_{i2}' H_{i2}^{-1} F_{i2} & 0 & 0 \\
\sum_{i=1}^{n} F_{i3}' H_{i3}^{-1} G_{i2} & \sum_{i=1}^{n} F_{i3}' H_{i3}^{-1} G_{i3} & \sum_{i=1}^{n} F_{i3}' H_{i3}^{-1} F_{i3} & 0 \\
\sum_{i=1}^{n} F_{i4}' H_{i4}^{-1} G_{i4} & 0 & \sum_{i=1}^{n} F_{i4}' H_{i4}^{-1} G_{i6} & \sum_{i=1}^{n} F_{i4}' H_{i4}^{-1} F_{i4}
\end{pmatrix},
$$

with

$$G_{i2} = \partial g(\mu, \varsigma^2, \lambda^2)/\partial \mu,$$

$$G_{i3} = \partial g(\mu, \varsigma^2, \lambda^2)/\partial \varsigma^2,$$

$$G_{i4} = \partial h(\mu, \lambda^2, \rho)/\partial \mu,$$

$$G_{i6} = \partial h(\mu, \lambda^2, \rho)/\partial \lambda^2,$$

and

$$
\Sigma = \begin{pmatrix}
A_{11} & A_{12} & A_{13} & A_{14} \\
A_{21} & A_{22} & A_{23} & A_{24} \\
A_{31} & A_{32} & A_{33} & A_{34} \\
A_{41} & A_{42} & A_{43} & A_{44}
\end{pmatrix},
$$

with

$$A_{11} = \sum_{i=1}^{n} F_{i1}' H_{i1}^{-1} (Q_{i1} - \hat{\mu})(Q_{i1} - \hat{\mu})' H_{i1}^{-1} F_{i1},$$

$$A_{12} = \sum_{i=1}^{n} F_{i1}' H_{i1}^{-1} (Q_{i1} - \hat{\mu})(Q_{i2} - \hat{\varsigma}^2)' H_{i2}^{-1} F_{i2},$$

$$A_{13} = \sum_{i=1}^{n} F_{i1}' H_{i1}^{-1} (Q_{i1} - \hat{\mu})(Q_{i3} - \hat{g}(\mu, \varsigma^2, \lambda^2))' H_{i3}^{-1} F_{i3},$$

$$A_{14} = \sum_{i=1}^{n} F_{i1}' H_{i1}^{-1} (Q_{i1} - \hat{\mu})(Q_{i4} - \hat{h}(\mu, \lambda^2, \rho))' H_{i4}^{-1} F_{i4},$$

$$A_{22} = \sum_{i=1}^{n} F'_{i2} H_{i1}^{-1} (Q_{i2} - \hat{\varsigma}^2)(Q_{i2} - \hat{\varsigma}^2)' H_{i2}^{-1} F_{i2},$$

$$A_{23} = \sum_{i=1}^{n} F'_{i2} H_{i2}^{-1} (Q_{i2} - \hat{\varsigma}^2)(Q_{i3} - \hat{g}(\mu, \varsigma^2, \lambda^2))' H_{i3}^{-1} F_{i3},$$

$$A_{24} = \sum_{i=1}^{n} F'_{i2} H_{i2}^{-1} (Q_{i2} - \hat{\varsigma}^2)(Q_{i4} - \hat{h}(\mu, \lambda^2, \rho))' H_{i4}^{-1} F_{i4},$$

$$A_{33} = \sum_{i=1}^{n} F'_{i3} H_{i3}^{-1} (Q_{i3} - \hat{g}(\mu, \varsigma^2, \lambda^2))(Q_{i3} - \hat{g}(\mu, \varsigma^2, \lambda^2))' H_{i3}^{-1} F_{i3},$$

$$A_{34} = \sum_{i=1}^{n} F'_{i3} H_{i3}^{-1} (Q_{i3} - \hat{g}(\mu, \varsigma^2, \lambda^2))(Q_{i4} - \hat{h}(\mu, \lambda^2, \rho))' H_{i4}^{-1} F_{i4},$$

$$A_{44} = \sum_{i=1}^{n} F'_{i4} H_{i4}^{-1} (Q_{i4} - \hat{h}(\mu, \lambda_e^2, \rho))(Q_{i4} - \hat{h}(\mu, \lambda^2, \rho))' H_{i4}^{-1} F_{i4},$$

and $A_{21} = A'_{12}$, $A_{31} = A'_{13}$, $A_{41} = A'_{14}$, $A_{32} = A'_{23}$, $A_{42} = A'_{24}$, $A_{43} = A'_{34}$.

Now we have obtained the estimates and variance–covariance matrix for all parameters. We can then use the delta method to obtain the estimates and their variances for all indices that are functions of these parameters.

6.4 Comparison of Total-Rater Agreement with Intrarater Precision: Total–Intra Ratio

For evaluating the type of individual agreement mentioned in the second paragraph of this chapter, it is natural to use TIR, the ratio of $\mathrm{MSD}_{\mathrm{total}}$ and $\mathrm{MSD}_{\mathrm{intra}}$, to assess the noninferiority between measurements from different raters to an intrarater precision. More generally, the comparison can be based on selected multiple pairs of $\mathrm{MSD}_{\mathrm{total}_{jj'}}$ relative to selected multiple $\mathrm{MSD}_{\mathrm{intra}_j}$. Selected raters for $\mathrm{MSD}_{\mathrm{total}}$ can also be selected for $\mathrm{MSD}_{\mathrm{intra}}$, and hence raters in the numerator and raters in the denominator are not mutually exclusive. This approach allows substantial flexibility in making comparisons between chosen test raters and chosen reference raters. In addition, it is not required to select a reference rater when none is available. For example, when $k = 2$, we can evaluate deviations among individual values of test and reference raters relative to the deviation within the reference raters, $\mathrm{MSD}_{\mathrm{total}_{T,R}}/\mathrm{MSD}_{\mathrm{intra}_R}$, or relative to that within both test and reference raters, $\mathrm{MSD}_{\mathrm{total}_{T,R}}/\mathrm{MSD}_{\mathrm{intra}_{T,R}}$. When $k = 3$ with one of the raters being the reference rater, we can evaluate $\mathrm{MSD}_{\mathrm{total}_{T_1,R}}/\mathrm{MSD}_{\mathrm{intra}_R}$, $\mathrm{MSD}_{\mathrm{total}_{T_2,R}}/\mathrm{MSD}_{\mathrm{intra}_R}$, or $\mathrm{MSD}_{\mathrm{total}_{T_1 T_2,R}}/\mathrm{MSD}_{\mathrm{intra}_R}$. When $k = 3$ and none is the reference, we can evaluate $\text{MSD}_{\text{total}}/\text{MSD}_{\text{intra}}$, $\text{MSD}_{\text{total}_{1,2}}/\text{MSD}_{\text{intra}_{1,2}}$, $\text{MSD}_{\text{total}_{1,3}}/\text{MSD}_{\text{intra}_{1,3}}$, and $\text{MSD}_{\text{total}_{2,3}}/\text{MSD}_{\text{intra}_{2,3}}$. In the following, we will discuss cases with at least one reference rater and those in which no reference rater is available. Here $\text{MSD}_{\text{intra}_{T,R}}$ is the average of $\text{MSD}_{\text{intra}_T}$ and $\text{MSD}_{\text{intra}_R}$, while $\text{MSD}_{\text{total}_{T_1 T_2,R}}$ is the average of $\text{MSD}_{\text{total}_{T_1,R}}$ and $\text{MSD}_{\text{total}_{T_2,R}}$.

6.4.1 When One or Multiple References Exist

In the case of one or multiple references, we select a set of test raters and a set of reference raters out of the total of k raters. Suppose we are interested in evaluating t different test raters, $1 \leqslant t \leqslant k$, with respect to r reference raters, $1 \leqslant r \leqslant k$, $2 \leqslant t + r \leqslant 2k$. Test raters are indexed by j, $j = 1, \ldots, t$, and reference raters are indexed by j', $j' = 1, \ldots, r$. The individual differences between selected sets of test and reference raters are evaluated by the average of the pairwise total mean squared deviation, $\text{MSD}_{\text{total}_{T,R}}$. The intrarater precision is evaluated by the average of intra mean squared deviation of selected r reference raters. The ratio, TIR_R is used to assess the individual agreement:

$$\zeta_R = \frac{\varepsilon_{\text{total}_{T,R}}^2}{\varepsilon_{\text{intra}_R}^2} = \frac{\sum_{j=1}^t \sum_{j'=1}^r E(y_{ijl} - y_{ij'l'})^2 / tr}{\sum_{j'=1}^r E(y_{ij'l} - y_{ij'l'})^2 / r}$$

$$= \frac{\frac{1}{tr} \sum_{j'=1}^r \sum_{j=1}^t (\mu_j - \mu_j')^2 + \frac{1}{t} \sum_{j=1}^t \varsigma_j^2 + \frac{1}{r} \sum_{j'=1}^r \varsigma_{j'}^2}{\frac{2}{r} \sum_{j'=1}^r \varsigma_{j'}^2}$$

$$+ \frac{\frac{1}{t} \sum_{j=1}^t \lambda_j^2 + \frac{1}{r} \sum_{j'=1}^r \lambda_{j'}^2 - \frac{1}{tr} \sum_{j'=1}^r \sum_{j=1}^t \rho_{jj'} \lambda_j \lambda_j'}{\frac{2}{r} \sum_{j'=1}^r \varsigma_{j'}^2}. \qquad (6.22)$$

Theoretically, TIR_R cannot be less than 1. However, its estimate, $\widehat{\text{TIR}}_R$, can be less than one due to random error. When $\text{TIR}_R = 1$, total-rater agreement is exactly the same as intra-reference-rater agreement. Higher values of TIR_R indicate worse individual agreement. The cause of disagreement could be due to (1) difference between the means μ_T of the test raters and means μ_R of the reference raters, (2) difference between the error variance ς_T^2 of the test raters and the error variance ς_R^2 of the reference raters, (3) the subject by rater interaction: $\sigma_D^2 = \text{var}(\mu_{iT} - \mu_{iR}) = \lambda_T^2 + \lambda_R^2 - 2\rho_{T,R}\lambda_T\lambda_R$.

6.4.1.1 When No Specific Reference Exists

When there is no specific reference rater, we can select t test raters $\text{MSD}_{\text{total}}$ relative to their $\text{MSD}_{\text{intra}}$. The individual difference is evaluated by the average of the pairwise total mean squared deviation $\text{MSD}_{\text{total}_T}$ among t selected test raters,

$2 \leqslant t \leqslant k$. We then use the average of $\text{MSD}_{\text{intra}}$ of all test raters in the denominator. The total–intra ratio without a specific reference, TIR_{all}, is expressed as

$$
\begin{aligned}
\zeta_{\text{all}} &= \frac{\varepsilon_{\text{total}}^2}{\varepsilon_{\text{intra}}^2} \\
&= \frac{2\sum_{j=1}^{t-1}\sum_{j'=j+1}^{t} E(y_{ijl} - y_{ij'l'})^2 / t(t-1)}{\sum_{j=1}^{t} E(y_{ijl} - y_{ijl'})^2 / t} \\
&= \frac{2\sum_{j=1}^{t-1}\sum_{j'=j+1}^{t}\left((\mu_j - \mu_{j'})^2 + \varsigma_j^2 + \varsigma_{j'}^2 + \lambda_j^2 + \lambda_{j'}^2 - \rho_{jj'}\lambda_j\lambda_{j'}\right)/t(t-1)}{2\sum_{j=1}^{t}\varsigma_j^2}.
\end{aligned}
$$

(6.23)

When no specific reference exists, TIR_{all} varies between 1 and ∞.

6.4.2   Comparison to FDA's Individual Bioequivalence with Relative Scale

In the case of two raters with one of them treated as a reference, i.e., $k = 2$, $t = 1$, and $r = 1$, TIR_R degenerates to FDA's method for evaluating individual bioequivalence under the relative scale. Following FDA guidelines on individual bioequivalence, the agreement of test and reference compounds can be assessed relative to the agreement of within-reference compound. Let y_{iTl} and $y_{iRl'}$ be the lth and l'th reading on subject i from test compound (T) and reference compound (R), respectively. Then the individual bioequivalence criterion (IBC) is defined by FDA as

$$
\text{IBC} = \frac{E(y_{iTl} - y_{iRl'})^2 - E(y_{iRl} - y_{iRl'})^2}{E(y_{iRl} - y_{iRl'})^2 / 2}.
$$

(6.24)

This FDA approach is primarily based on the approach proposed by Sheiner (1992), which uses a normal linear mixed model estimated by REML.

By our definition, TIR_R is expressed as

$$
\begin{aligned}
\zeta_R &= \frac{\varepsilon_{\text{total}T,R}^2}{\varepsilon_{\text{intra}R}^2} \\
&= \frac{E(y_{iTl} - y_{iRl'})^2}{E(y_{iRl} - y_{iRl'})^2} \\
&= \frac{\text{IBC}}{2} + 1.
\end{aligned}
$$

(6.25)

Note that FDA also uses a constant scale when $\text{MSD}_{\text{intra}R}$ is small. In addition, it requires that the ratio of the geometric mean of the test and reference compound lie between 0.8 and 1.25. We will discuss in detail later, in Example 6.7.4, the topic of individual bioequivalence. In that example, we will add meaningful interpretations to the FDA's criteria by making use of the information presented in Chapters 5 and 6.

6.4.3   Comparison to Coefficient of Individual Agreement

Barnhart, Kosinski, and Haber (2007) proposed the coefficient of individual agreement (CIA) for assessing individual agreement. When there are t test raters and r reference raters, CIA with reference is defined as

$$\text{CIA}_R = \frac{\varepsilon_{\text{intra}R}^2}{\varepsilon_{\text{total}T,R}^2} = \frac{\sum_{j=1}^{r} E(y_{ij'l} - y_{ij'l'})^2/r}{\sum_{j=1}^{t}\sum_{j'=1}^{r} E(y_{ijl} - y_{ij'l'})^2/tr}. \tag{6.26}$$

When no reference is available, CIA without reference is defined as

$$\text{CIA}_{\text{all}} = \frac{\varepsilon_{\text{intra}}^2}{\varepsilon_{\text{total}}^2} = \frac{\sum_{j=1}^{t} E(y_{ijl} - y_{ijl'})^2/t}{2\sum_{j=1}^{t-1}\sum_{j'=j+1}^{t} E(y_{ijl} - y_{ij'l'})^2/t(t-1)}. \tag{6.27}$$

Compare these CIAs to TIR_R and TIR_{all} in (6.22) and (6.23), respectively: CIA is the reciprocal of TIR.

Basically, IBC, CIA, and TIR are the same indices. The differences are in the estimation approaches, in that the IBC is ML-based, CIA is method-of-moments-based with bootstrapping for statistical inference, and TIR is GEE-based. CIA and TIR are extended for multiple raters, while IBC is limited to two raters only.

6.4.4   Estimation and Asymptotic Normality

Recall that we have obtained estimates of all parameters as well as their variance–covariance matrix via GEE methodology in Section 6.3. These GEE estimates of parameters turn out to be moment estimates. Since TIR is a function of the parameters in the model, the method of moments is used to estimate TIR, and the delta method is used for the statistical inference.

When the reference exists, the TIR_R estimate can be obtained as

$$\hat{\zeta}_R = \frac{\frac{1}{tr}\sum_{j'=1}^{r}\sum_{j=1}^{t}(\hat{\mu}_j - \hat{\mu}_j')^2 + \frac{1}{t}\sum_{j=1}^{t}\hat{s}_j^2 + \frac{1}{r}\sum_{j'=1}^{r}\hat{s}_{j'}^2}{\frac{2}{r}\sum_{j'=1}^{r}\hat{s}_r^2}$$

$$+ \frac{\frac{1}{t}\sum_{j=1}^{t}\hat{\lambda}_j^2 + \frac{1}{r}\sum_{j'=1}^{r}\hat{\lambda}_r^2 - \frac{1}{tr}\sum_{j'=1}^{r}\sum_{j=1}^{t}\hat{\rho}_{jj'}\hat{\lambda}_j\hat{\lambda}_j'}{\frac{2}{r}\sum_{j'=1}^{r}\hat{s}_r^2}. \tag{6.28}$$

The estimate of the log transformed TIR_R, $W_T = \ln(\hat{\zeta}_R)$, has an asymptotic normal distribution with mean $\ln(\zeta_R)$ and variance

$$\sigma_{W_T}^2 = \frac{1}{n} d_T' \Sigma_C d_T, \tag{6.29}$$

where $W_T = \ln(\hat{\zeta}_R) = g(m)$,

$$m = (\hat{\mu}, \hat{\varsigma}^2, \hat{\lambda}^2, \hat{\rho})' = (m_1, m_2, m_3, m_4)',$$

$\Sigma_C = nV(\hat{\theta})$ from (6.21), which is the variance–covariance matrix for the parameter estimates,

$$d_T = (d_\mu, d_{\varsigma^2}, d_{\lambda^2}, d_\rho)'$$
$$= \left(\frac{\partial g(m)}{\partial m_1}\bigg|_{m=\Theta}, \frac{\partial g(m)}{\partial m_2}\bigg|_{m=\Theta}, \frac{\partial g(m)}{\partial m_3}\bigg|_{m=\Theta}, \frac{\partial g(m)}{\partial m_4}\bigg|_{m=\Theta} \right)',$$

and $\Theta = (\mu, \varsigma^2, \lambda^2, \rho)'$. We use the correction factor $\frac{n}{n-6}$ for the variance in (6.29) because it has been shown to have less bias in the simulation studies.

The elements of d_T are computed as follows: If the jth rater is selected as the test rater and the j'th rater is selected as the reference rater, then:

- The jth element of d_T is $\frac{1}{\zeta_R}\left(\frac{\sum_{j'=1}^r (\mu_j - \mu_{j'})}{t\sum_{j'=1}^r s_{j'}^2} \right)$.
- The j'th element of d_T is $\frac{1}{\zeta_R}\left(\frac{-\sum_{j=1}^t (\mu_j - \mu_{j'})}{t\sum_{j'=1}^r s_{j'}^2} \right)$.
- The $(k+j)$th element of d_T is $\frac{1}{\zeta_R}\left(\frac{r}{t\sum_{j'=1}^r s_{j'}^2} \right)$.
- The $(k+j')$th element of d_T is
$$\frac{1}{\zeta_R}\left(\frac{\frac{1}{tr}\sum_{j'=1}^r \sum_{j=1}^t (\mu_j - \mu_j')^2 + \frac{1}{t}\sum_{j=1}^t s_j^2 + \frac{1}{t}\sum_{j=1}^t \lambda_j^2 + \frac{1}{r}\sum_{j'=1}^r \lambda_{j'}^2 - \frac{1}{tr}\sum_{j'=1}^r \sum_{j=1}^t \rho_{jj'}\lambda_j\lambda_j'}{\frac{2}{r}(\sum_{j'=1}^r s_{j'}^2)^2} \right).$$
- The $(2k+j)$th element of d_T is $\frac{1}{\zeta_R}\left(\frac{r - \sum_{j'=1}^r \rho_{jj'} \frac{\lambda_{j'}}{\lambda_j}}{2t \sum_{j'=1}^r s_{j'}^2} \right)$.
- The $(2k+j')$th element of d_T is $\frac{1}{\zeta_R}\left(\frac{t - \sum_{j=1}^t \rho_{jj'} \frac{\lambda_j}{\lambda_{j'}}}{2t \sum_{j'=1}^r s_{j'}^2} \right)$.
- When $j < j'$, the $\left(3k + \frac{(2k-j)(j-1)}{2} + (j' - j) \right)$th element of d_T is $\frac{1}{\zeta_R}\left(\frac{-\lambda_j \lambda_{j'}}{t\sum_{j'=1}^r s_{j'}^2} \right)$.
- When $j > j'$, the $\left(3k + \frac{(2k-j')(j'-1)}{2} + (j - j') \right)$th element of d_T is $\frac{1}{\zeta_R}\left(\frac{-\lambda_j \lambda_{j'}}{t\sum_{j'=1}^r s_{j'}^2} \right)$.
- All other elements in d_T are zero.

The log transformed TIR$_R$ estimate approaches normality rapidly and can efficiently bound the confidence interval within 0 to ∞. The confidence limit for TIR$_R$ is computed based on the log-transformed TIR estimates, $W_T = \ln(\hat{\zeta}_R)$. The antilog transformation is performed on the confidence limit of W_T to obtain the actual confidence limit for TIR$_R$. Individual agreement is established when the confidence limit is smaller than the prespecified criterion, say, 2.25.

When no reference raters exist and the average of all error variances of the test raters are used in the denominator, the TIR$_{all}$ estimate is given by

$$\hat{\zeta}_{all} = \frac{2\sum_{j=1}^{t-1}\sum_{j'=j+1}^{t}\left((\hat{\mu}_j - \hat{\mu}_{j'})^2 + \hat{s}_j^2 + \hat{s}_{j'}^2 + \hat{\lambda}_j^2 + \hat{\lambda}_{j'}^2 - \hat{\rho}_{jj'}\hat{\lambda}_j\hat{\lambda}_{j'}\right)/t(t-1)}{2\sum_{j=1}^{t}\hat{s}_j^2}.$$

(6.30)

The statistical inference for TIR$_{all}$ can be obtained in the same way as when there are reference raters. For the purpose of statistical inference, the parameters in the variances of estimates presented above can be replaced by their sample counterparts that are consistent estimators.

6.5   Comparison of Intrarater Precision Among Selected Raters: Intra–Intra Ratio

In the medical-device environment, we are often interested in whether the within-device precision of a newly developed device can be better than, equal to, or worse than that of the within-device precision for the old device. In our approach, we can select any set of test raters as well as any set of reference raters out of the total of k raters, and the intraprecision of the selected set of test raters is compared to that of a selected set of reference raters. Here, the selected two sets of test and reference raters are mutually exclusive. For example, when $k = 2$, we can evaluate only MSD$_{intra_1}$/MSD$_{intra_2}$ and MSD$_{intra_2}$/MSD$_{intra_1}$. When $k = 3$, we can evaluate MSD$_{intra_{1,2}}$/MSD$_{intra_3}$, MSD$_{intra_{1,3}}$/MSD$_{intra_2}$, and MSD$_{intra_{2,3}}$/MSD$_{intra_1}$. When we select t test raters and r reference raters, $1 \leqslant t \leqslant k$, $1 \leqslant r \leqslant k$, $2 \leqslant t + r \leqslant 2k$, the IIR can be expressed as

$$\psi = \frac{\varepsilon_{intra_T}^2}{\varepsilon_{intra_R}^2}$$

$$= \frac{\sum_{j=1}^{t} E(y_{ijl} - y_{ijl'})^2/t}{\sum_{j'=1}^{r} E(y_{ij'l} - y_{ij'l'})^2/r}$$

$$= \frac{\sum_{j=1}^{t} s_j^2/t}{\sum_{j'=1}^{r} s_{j'}^2/r},$$

(6.31)

where $\varepsilon_{intra_T}^2$ and $\varepsilon_{intra_R}^2$ denote MSD$_{intra}$ among selected test and reference raters, respectively.

IIR less than 1 indicates better overall precision for test raters than reference raters. For statistical inference, we would construct a two-sided $100(1 - \alpha/2)\%$ confidence interval for IIR, and claim superiority or inferiority if the upper or lower limit is less than or greater than 1.0. On the other hand, if we intend to examine whether precision of the test and reference raters are equal, we would claim equivalence if the confidence interval is bounded by a prespecified clinically relevant interval.

6.5.1 Estimation and Asymptotic Normality

Similarly as for TIR, the estimate for IIR is obtained via the method of moments by GEE methodology, and the related statistical inference is obtained via the delta method. The IIR estimate is given by

$$\hat{\psi} = \frac{\sum_{j=1}^{t} \hat{\varsigma}_j^2 / t}{\sum_{j'=1}^{r} \hat{\varsigma}_{j'}^2 / r}. \tag{6.32}$$

The estimate of the log-transformed IIR, $W_I = \ln(\hat{\psi})$, has an asymptotic normal distribution with mean $\ln(\hat{\psi})$ and variance

$$\sigma_{W_I}^2 = \frac{1}{n} d_I' \Sigma_C d_I, \tag{6.33}$$

where $W_I = \ln(\hat{\psi}) = g(m)$,

$$m = (0, \hat{\varsigma}^2, 0, 0)$$
$$= (0, m_2, 0, 0),$$

and

$$d_I = (0, d_{\varsigma^2}, 0, 0)$$
$$= \left(0, \left.\frac{\partial g(m)}{\partial m_2}\right|_{m=\Theta}, 0, 0\right).$$

We use the correction factor $\frac{n}{n-6}$ for the variance in (6.33) because it has been shown to have less bias in simulation studies.

The elements of d_I are computed as follows: If the jth rater is selected as the test rater and the j'th rater is selected as the reference rater, then:

• The $(k + j)$th element of d_I is $\frac{1}{\hat{\psi}}\left(\frac{r}{t\sum_{j'=1}^{r} \varsigma_{j'}^2}\right)$.

- The $(k + j')$th element of $\boldsymbol{d}_I$ is $-\frac{1}{\psi}\left(\frac{r\sum_{j=1}^{t} s_j^2}{t(\sum_{j'=1}^{r} s_{j'}^2)^2}\right)$.
- All other elements in $\boldsymbol{d}_I$ are zero. The value of the elements in $\boldsymbol{d}_I$ correspond to the selection of the test and reference raters.

The transformed IIR estimate approaches normality rapidly, and the confidence interval for IIR is bounded within 0 to ∞. The confidence limit for IIR is computed based on the log-transformed IIR estimates, $W_I = \ln(\hat{\psi})$. The antilog transformation is performed on the confidence limit of W_T to obtain the actual confidence limit for IIR. Better precision for test raters than reference raters is determined when the confidence limit is less than 1. For the purpose of statistical inference, the parameters in the variances of estimates presented above can be replaced by their sample counterparts that are consistent estimators.

6.6 Summary of Simulation Results

Simulation studies were conducted to assess the performance of the GEE methodology for estimation and inference of TIR and IIR for three different types of data: binary, ordinal, and normal when $k = 2$ and $m = 2$. A total of 1,000 random samples was generated with size of 40 for normal data and 80 for binary and ordinal data. For each type of data, the simulation was designed to evaluate both the significance level and the power.

For TIR with criterion 2.25, we generated data with TIR $= 2.25$ to assess the significance level. We generated data with TIR $= 1.25$ for normal data and TIR $= 1.33$ for categorical data to assess the power. For IIR, we generated data with IIR $= 1$ to assess the significance level. We generated data with IIR $= 0.5$ to assess the power. The estimates and their standard errors correspond to their theoretical values very well. The coverage probability to assess a significant level is close to 0.05, and powers vary from 0.45 to 0.80. The details of the simulation results can be seen in Lin, Hedayat, and Tang (2012).

6.7 Examples

6.7.1 Example 1: TIR and IIR for an Automatic Blood Pressure Meter

In Example 5.9.4, we examined the intraagreement, interagreement, and total agreement of one semiautomated blood pressure meter (S) and the gold standard sphygmomanometer measured by two medical staffs (J and R) with triplicate measurements by each. Figs. 5.4–5.9 show that it is informative to investigate the

TIR of $\text{MSD}_{\text{total}_{S,JR}}$ relative to $\text{MSD}_{\text{intra}_{JR}}$, and the IIR of $\text{MSD}_{\text{intra}_S}$ relative to $\text{MSD}_{\text{intra}_{JR}}$, because readings from J and R were precisely interchangeable, as evident in Fig. 5.4. Here, data were assumed to have a proportional error structure and were analyzed with a log transformation to the data.

We now evaluate the TIR of $\text{MSD}_{\text{total}_{S,JR}}$ relative to $\text{MSD}_{\text{intra}_{JR}}$. The TIR_R of S relative to J and R is estimated to be 7.06 with the one-tailed 95% upper confidence limit 10.45, which is much greater than any clinically relevant criterion, indicating that S is not interchangeable with J and R. The IIR estimate is 1.57 with the two-sided 95% confidence interval (1.05, 2.33). The lower limit is greater than 1, indicating that the precision of S is significantly inferior to the precision of J and R.

The results imply that the automatic blood pressure monitor does not have good individual agreement with sphygmomanometers used by medical staff and that the precision of the automatic blood pressure apparatus is significantly inferior to that of a sphygmomanometer. Therefore, we would not want to replace a sphygmomanometer with the studied automatic machine used in this study.

6.7.2 Example 2: Nasal Bone Image Assessment by Ultrasound Scan

In Example 5.9.3, we examined the intraagreement, interagreement, and total agreement of nasal bone (NB) image to be used for population screening in an Australian obstetric population at 11–13 weeks gestation. Recall that three raters assessed each of the 400 images twice ($k = 3, m = 2$).

Tables 5.7–5.12 present the frequency tables among three raters and their duplicate readings. It appears that within raters has a better agreement than between raters, and it is informative to investigate the TIR without any reference examiners. This TIR is estimated to be 1.56 with 95% upper confidence limit 2.38. We can accept the between-raters agreement if we consider using the criterion 2.5.

6.7.3 Example 3: Validation of the Determination of Glycine on a Spectrophotometer System

The data set for this example is from Baxter Healthcare Corporation. Spectrophotometer systems are used to measure the spectrophotometric determination of glycine. In this example, we examine the agreement between two different systems, S_1 and S_2, with duplicate measurements on each of 38 samples. Figures 6.1–6.3 present the graphs among readings between the two systems and their replicated measurements. There are no reference instruments here, and hence it would be informative to investigate the TIR and IIR without reference. Here, data were assumed to have a constant error structure.

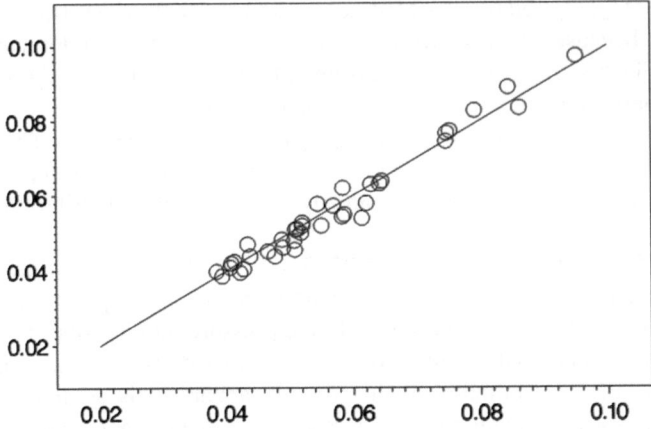

Fig. 6.1 S_1 reading 1 vs. S_1 reading 2

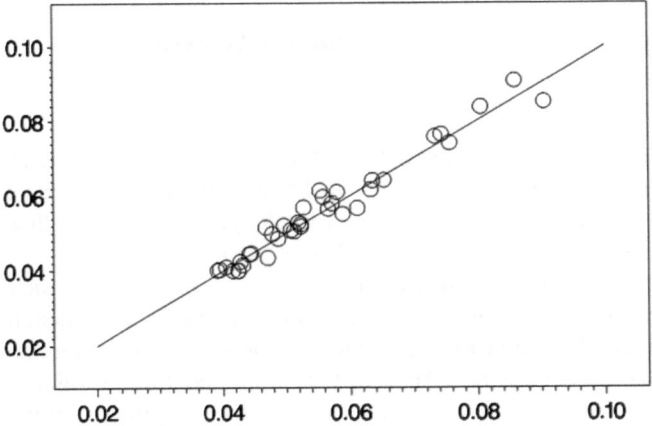

Fig. 6.2 S_2 reading 1 vs. S_2 reading 2

The TIR$_{all}$ of MSD$_{total}$ relative to MSD$_{intra}$ is estimated to be 0.699 with the 95% upper confidence limit of 0.817, which is within any clinically relevant criterion for claiming individual agreement. The IIR of MSD$_{intra_{S_1}}$ relative to MSD$_{intra_{S_2}}$ is estimated to be 1.010 with 95% confidence interval (0.752, 1.357), which indicates that the precisions of the two systems are not statistically different. We would claim equivalence if we considered precision deviation of less than 40% clinically acceptable. Based on the result of TIR and IIR, we conclude that the two systems can be used interchangeably.

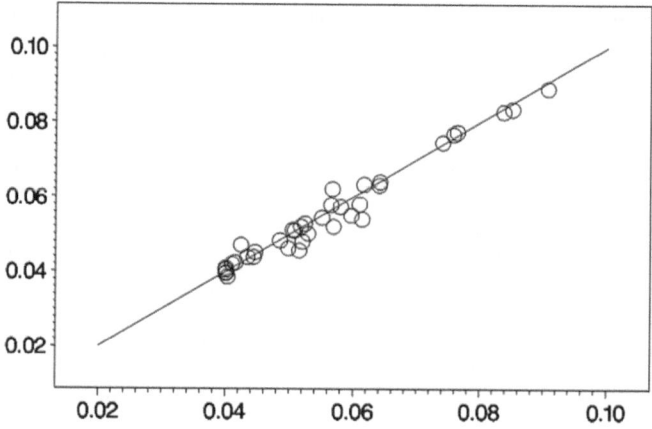

Fig. 6.3 S_1 reading 1 vs. S_2 reading 1

6.7.4 Example 4: Individual Bioequivalence

To study individual bioequivalence, we downloaded a set of data with a four-period–four-sequence crossover design from the FDA web site. We do not know the product name or the manufacturer. There were 40 healthy volunteers in the study. The design of the study is given in the table below, where T represents the test compound and R represents the reference compound.

Seq.	Period			
	1	2	3	4
1	T	T	R	R
2	R	R	T	T
3	R	T	T	R
4	T	R	R	T

For simplicity, we study only the area under the curve (AUC), and we assume that the data have proportional error structure. To save space, we do not list the original data, but we will shortly present the condensed data. We first tested the period and sequence effects using the mixed effect model on log-transformed AUC, and found that there was no evidence of these two effects with $p > 0.5$. Therefore, we list the data in the format T1, T2, R1, and R2, representing periods 1 and 2 of the test and reference compounds, as shown in Table 6.1. Note that subject 16 had missing data, and subject numbers 906, 908, 921, and 932 were recruited to replace the four dropout subjects with numbers 6, 8, 21, and 32.

Table 6.1 Data listing

Subject	T1	T2	R1	R2
1	13	16.5	16.2	9.35
2	27.3	58.6	23	33.3
3	8.11	9.28	63.1	4.76
4	4.62	5.88	6.15	6.78
5	3.77	6	6.9	6.72
7	20	17.4	28.6	26.1
9	15.6	14.1	8.95	17.6
10	5.66	4.56	3.62	5.07
11	17.4	21.5	13.4	13.1
12	4.9	4.83	4.13	3.07
13	39.2	32.8	33.7	28.5
14	6.78	5.34	10.2	16.7
15	15.8	30.7	20.5	17.9
16	.	.	.	.
17	10.3	12.2	11.5	9.58
18	31.6	36.7	22.4	25.2
19	11.5	10.8	18.3	32.4
20	8.29	9.31	6.26	10.5
22	6.17	7.86	7.98	4.54
23	44.3	22.2	35	20.3
24	22.5	60.2	31.5	15.1
25	9.63	5.76	8.34	3.85
26	5.35	5.09	5.33	4.78
27	8.95	4.82	4.18	10.2
28	5.1	9.68	7.08	9.67
29	6.53	3.7	7.88	6.23
30	3.05	5.12	2.29	5.48
31	7.41	5.98	5.35	5.79
33	0.82	1.91	2.63	0.83
34	3.43	5.34	6.91	7.66
35	8.58	8.48	5.71	6.64
36	8.34	9.12	6.9	12.2
37	15.6	15.8	12.3	14.6
38	3.65	5.63	8.81	13.6
39	6.72	5.58	11.6	7.38
40	20.4	20.6	19.3	20.5
906	7.11	5.41	6.55	4.55
908	11.9	12.6	8.68	6.48
921	14.7	13.6	12.9	24.5
932	10.6	20.8	9.09	13.7

T1: AUC value of the period-1 test compound
T2: AUC value of the period-2 test compound
R1: AUC value of the period-1 reference compound
R2: AUC value of the period-2 reference compound

The criteria for claiming individual bioequivalence under the FDA guidance are

1. Reference Scale:

$$\text{IBC} = \frac{E(y_{iTl} - y_{iTl'})^2 - E(y_{iRl} - y_{iRl'})^2}{E(y_{iRl} - y_{iRl'})^2/2}$$

$$= \frac{(\mu_T - \mu_R)^2 + \sigma_D^2 + \sigma_{WT}^2 - \sigma_{WR}^2}{\sigma_{WR}^2} < 2.5, \text{ if } \sigma_{WR}^2 > \sigma_{W0}^2, \quad (6.34)$$

2. Constant Scale:

$$\frac{(\mu_T - \mu_R)^2 + \sigma_D^2 + \sigma_{WT}^2 - \sigma_{WR}^2}{\sigma_{W0}^2} < 2.5, \text{ if } \sigma_{WR}^2 \leqslant \sigma_{W0}^2, \quad (6.35)$$

where $\sigma_D^2 = \sigma_{BT}^2 + \sigma_{BR}^2 - \sigma_{RT}$, μ_T and μ_R are means, σ_{BT}^2 and σ_{BR}^2 are between-subject variances, σ_{WT}^2 and σ_{WR}^2 are within-subject variances, and σ_{RT} is the between-subject covariance, of the test and reference compounds, respectively. Finally, σ_{W0}^2 is the cutoff value based on the within-subject variance of the reference compound. The criterion 2.5 is the aggregate allowance of $\ln(\mu_T/\mu_R) = \ln(1.25)$, $\sigma_{WT}^2 - \sigma_{WR}^2 = 0.02$, $\sigma_D^2 = 0.03$, and $\sigma_{W0}^2 = 0.04$.

The FDA individual bioequivalence criteria have not been used widely, even by FDA staff, for the following reasons:

1. It is difficult for pharmacists and statisticians to understand the meaning of the criteria.
2. The concept is far more complicated than the average or population bioequivalence.
3. For statisticians, it is complicated to compute the confidence limits of the estimates of the relative and constant scales unless there is a program or macro ready at hand.
4. Perhaps the biggest problem is that there is a discontinuity region in using the reference and constant scales. That is, when the estimate of σ_{WR}^2 is near σ_{W0}^2 within a natural random fluctuation, there is a penalty to fall into the more strict constant scale criterion by chance.

Using our definitions provided in Chapters 5 and 6, we can much better interpret and understand these scales and provide tools (see Chapter 7) to do the analysis. See Section 6.4.1 for the relationship between TIR and the relative scale, while the relationship between TIR_R and the relative scale is given in (6.25). Using our definition, the relative scale criterion means that $\text{MSD}_{\text{total}_{T.R}}$ cannot be more than 2.25 of $\text{MSD}_{\text{intra}_R}$, or $\text{TIR}_R < 2.25$, with $(1 - \alpha)\%$ confidence.

We now examine the meaning of $\sigma_{W0}^2 = 0.04$. From (2.8) and (5.11), we can convert $\sigma_{W0}^2 = 0.04$ to $TDI_{0.8}$ of 0.3625 based on the log scale because $MSD_{intra_R} = 2\sigma_{W0}^2 = 0.08$. From (5.33), the $TDI\%_{0.8}$ becomes 43.7%. Therefore, $\sigma_{W0}^2 = 0.04$ means that 80% of the duplicate values of the reference compound are within 43.7% of each other. If $TDI\%_{0.8}$ is greater than 43.7%, we must use the reference scale according to (6.34) to ensure that the $MSD_{total_{T,R}}$ cannot be more than 2.25 of the MSD_{intra_R} with $100(1 - \alpha)\%$ confidence. Otherwise, we would use the constant scale.

We now examine the meaning of the constant scale. Note that $MSD_{total_{T,R}} = (\mu_T - \mu_R)^2 + \sigma_D^2$ under model (6.1). If we assume that $\sigma_{WT}^2 - \sigma_{WR}^2 = -0.02, 0,$ or 0.02, (6.35) under model (6.1) becomes $MSD_{total_{T,R}} < 0.12, 0.1,$ or 0.08, respectively. Again, when $\sigma_{W0}^2 = 0.04$, from (2.8), (5.26), and (5.33), the $TDI\%_{total(0.8)}$ becomes 55.9%, 50.0%, or 43.7%, respectively. This means that approximately 80% of the individual AUC values from the test compound cannot deviate more than 55.9%, 50.0%, or 43.7%, respectively, of the individual AUC values from the reference compound with $100(1 - \alpha)\%$ confidence. Compared to the $TDI\%_{intra(0.8)} = 43.7\%$ based on the cutoff value of $\sigma_{W0}^2 = 0.04$ of the intrareference compound, this constant scale criterion appears to be too stringent to meet.

Let us summarize the interpretation of the FDA individual bioequivalence criteria under model (6.1). If 80% of the duplicate values of the reference compound deviate more than 43.7% from each other, the $MSD_{total_{T,R}}$ cannot be more than 2.25 of the MSD_{intra_R} with $100(1 - \alpha)\%$ confidence. Otherwise, we must show that approximately 80% of the individual AUC values from the test compound cannot deviate more than 44% to 56% of the individual AUC values from the reference compound with $100(1 - \alpha)\%$ confidence. The conventional α is set at 0.05, one-tailed. It appears that there is room for improvement in redefining these criteria.

We now analyze the data in this example. We begin by calculating the $TDI\%_{0.8}$ between R1 and R2 using (2.8), assuming the proportional error case. Figure 6.4 shows the agreement plot of R1 and R2 in $\log_2$ scale. It is clear that there exists insufficient precision in the data, and that $TDI\%_{0.8} = 124.4\%$, which means that 80% of R1 and R2 pairs can deviate up to 124%. This is much higher than the 43.7% cutoff value, and therefore we would use the relative scale according to the FDA rules.

Figure 6.5 shows the agreement plot of T1 and R2 in $\log_2$ scale, representing the total agreement among the individual AUC values of the test and reference compounds. The precision of the data is slightly better than that between R1 and R2, but in general, T1 has lower AUC values than R2. The TIR_R is estimated to be 0.691, with upper one-sided 95% confidence limit 1.076, indicating that the $MSD_{total_{T,R}}$ is 1.076 of the MSD_{intra_R} with $100(1 - \alpha)\%$ confidence, which is much less than 2.25. Therefore, individual bioequivalence is accepted according to FDA's rules because of the large within-reference deviation.

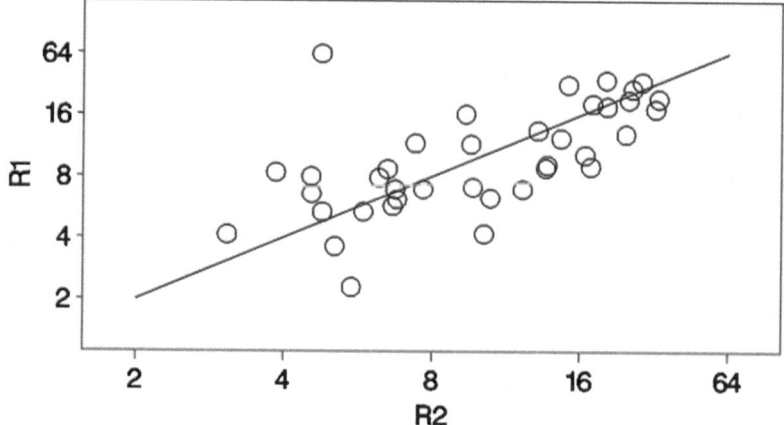

Fig. 6.4 Agreement between the duplicates of the reference compound

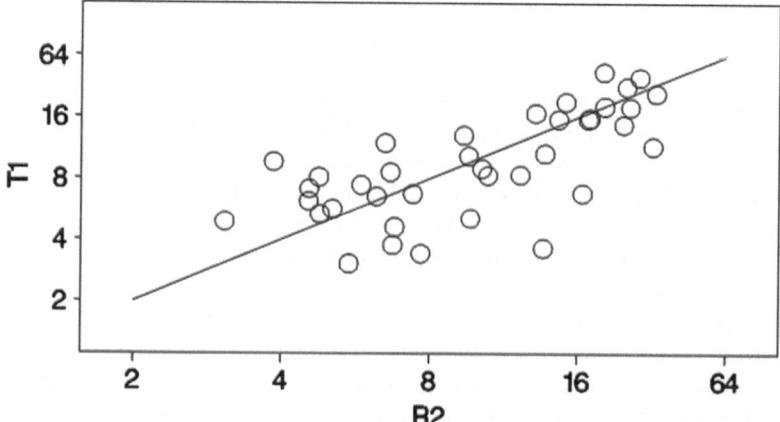

Fig. 6.5 Agreement between the first test compound reading and the second reference compound reading

Our statistical tools allow us to go one step further by examining the precision of the test compound relative to that of the reference compound, namely, the IIR of MSD_{intra_T} relative to MSD_{intra_R}. Figure 6.6 presents the agreement plot of T1 and T2 in $\log_2$ scale. The precision of the test compound appears tighter than that of the reference compound, as shown in Fig. 6.4. Using the statistical tools presented in Chapter 2, the $TDI\%_{0.8}$ of T1 and T2 is estimated to be 70.2%, with a 95% upper confidence limit of 90.3%. The IIR of MSD_{intra_T} relative to MSD_{intra_R} is estimated to be 0.432 with 95% confidence interval 0.168 to 1.115.

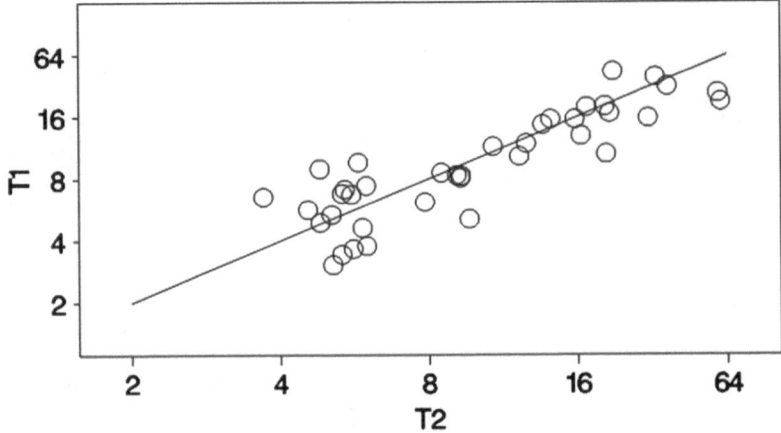

Fig. 6.6 Agreement between the duplicate of the test compound

6.8   Discussion

We have proposed two comparative agreement indices, TIR for interchangeability, and IIR for comparing precision measurements between multiple raters with multiple readings. TIR is a noninferiority assessment for determining whether several assays/raters can be used interchangeably, in terms of its ratio to the intrarater precision, for scenarios with and without a reference. IIR is an assessment for evaluating whether the precision of one or multiple raters is better than, equal to, or worse than that of other raters.

The approach we have proposed here is very general in a sense that any one or multiple raters can be selected as the test or reference raters. Users have substantial flexibility to make comparisons among any number of raters of interest. The FDA's method for evaluating individual bioequivalence under the relative scale becomes a special case of ours. The examples in this chapter show that our approaches have wide application in a variety of agreement studies.

We use GEE methodology because it is applicable to continuous, ordinal, and binary data. Lin, Hedayat, and Wu (2007) showed, in their simulation study when $m = 1$, that using ML or RML as proposed by Carrasco and Jover (2003) works well for ordinal data, but not for binary data. For ordinal data, we expect that the ML or RML approach would work well for the comparative model as well. The ML-based TIR and IIR should be an interesting topic for future research.

The bound for claiming individual agreement, by FDA's definition, is based on the upper limit $\text{TIR}_R = 2.25$. It is likely that the use of the 2.25 TIR criterion is too stringent for categorical data. These limits have been established based on experience with bioavailability data and with the average bioequivalence criterion. Users should carefully choose the boundary based on their specific clinical and historical assessments.

Our approach is based on balanced data. Further research is needed to investigate the case of unbalanced data with the possibility of covariate adjustments. For sample size and power calculations of TIR and IIR, the reader is referred to the general procedures outlined in Section 4.1.

There are relatively few references available regarding the topics in this chapter. Westlake (1976) proposed a modification of the conventional confidence interval method to obtain symmetric confidence intervals around 0.0 for bioequivalence trials. FDA guidance for industry (2001) introduced the statistical approach of evaluating individual bioequivalence between test and reference drugs, which was largely based on Sheiner (1992). Barnhart, Kosinski, and Haber (2007) proposed the concept of individual equivalence and extended FDA's approach of comparing two raters to the interchangeability among multiple raters. We have no knowledge of any publications regarding IIR.

Chapter 7
Workshop

In this chapter, we will walk through three examples using SAS macros and R functions. We will present the calling codes with in-depth explanations for each macro or function. For the first two examples, one with continuous data and one with categorical data, we will study from the basic models to more complicated models using the material presented in Chapters 2, 3, 5, and 6. We will also include an individual bioequivalence example using the material presented in Chapters 2, 5, and 6.

We have produced and made available three SAS macros: `Agreement`, `UnifiedAgreement`, and `TIR_IIR`. To run one of these macros, we need to download it from one of the following websites:

1. http://www.uic.edu/-hedayat/
2. http://mayoresearch.mayo.edu/biostat/sasmacros.cfm

These websites also contain most of the data used in this book.

An R package named `Agreement` is also available containing functions corresponding to the three SAS macros. These are also available for downloading from the above websites, or they can be installed directly from the comprehensive R archive network (CRAN).

7.1 Workshop for Continuous Data

In each of the following examples, we will first present the procedure based on the SAS macros, and then present the corresponding R functions.

7.1.1 The Basic Model (m=1)

Let us begin with the continuous data based on Example 5.9.1. This example is also presented in Example 2.8.1 using the basic model by taking the average of the duplicate readings for each of the HemoCue and Sigma assays. For Example 2.8.1, we execute the macro using the %agreement statement with parameters defined below:

- Dataset: the name of your data. We must avoid using dataset names of c, cc, t, tt, tb, and p, because these dataset names are used in this macro.
- Y: reading of the test assay or rater that will be shown on the vertical axis of the agreement plot.
- V_label: label for the vertical axis of the agreement plot.
- X: reading for the target assay or rater that will be shown on the horizontal axis of the agreement plot.
- H_label: label for the horizontal axis (target) of the agreement plot.
- Min: minimum of the plotting range.
- Max: maximum of the plotting range.
- By:

  - For the constant error structure, this is the increment of the plotting range. For example, by = 5.
  - For the proportional error structure, these are the log scale increments between min and max. For example, if the data range is from 1 to 60, then min = 1, max = 64, by = 2, 4, 8, 16, 32.

- Error: constant or proportional error structure.

  - error = const: the constant error structure. Here, TDI is expressed as an absolute difference with the same measurement unit as the original data.
  - error = prop: the proportional error structure. Here, TDI is expressed as a percent change. The natural log transformation to the data will be applied.

- CCC_a: a CCC allowance, which can be set as missing if there is no prespecified allowance.
- CP_a: a CP allowance that must be specified for computing TDI.
- TDI_a: a TDI allowance that must be specified for computing CP, and must be a percent value when error = prop is specified or an absolute difference when error = const is specified.
- Target: random or fixed.
- Alpha: $100(1 - \alpha)\%$ one-tailed confidence limit. The default is 0.05.
- Dec: significant digits after the decimal point printed for TDI when error = const is specified. The default is dec = 2.

The SAS macro `Agreement.sas` can be executed using the following code:

```
libname ex 'x:\xx\xxx\xxxx';

data e2_1;
  set ex.example2_1;
  HemocueAve=mean(of hemocue1 hemocue2);
  SigmaAve=mean(of sigma1 sigma2);
run;

goptions reset=all ftext=swiss htext=1;

%inc 'x:\xxx\xxx\xxx\Agreement.sas';

ods rtf file='x:\xxx\xxx\....\xxx.rtf' style=styles.TablesRTF;

%agreement(dataset=e2_1, y=HemocueAve, x=SigmaAve, V_label=
HemaCue, H_label=Sigma, min=0, max=2000, by=250, CCC_a=0.9775,
CP_a=0.9, TDI_a=150, error=const, target=random, dec=1,
alpha=0.05);

ods rtf close;
```

In the code above, the dataset name is `example2_1` in the `libname ex` statement. The variables `hemocue1` and `hemocue2` are duplicate values of HemoCue. The variables `sigma1` and `sigma2` are duplicate values of Sigma. The data step is used to compute the average of duplicates for each assay, namely `HemoCueAve` as the test assay that will be shown in the vertical axis, and `SigmaAve` as the target assay that will be shown in the horizontal axis. The `ods rtf file` statement is used to define the output destination where the agreement table and plot are located. The use of `style=styles.TablesRTF` is a predefined style for the table layout, which is not needed if we use the default layout. The use of `goptions reset=all ftext=swiss htext=1` is to define our desirable graphic options for the agreement plot. We then give the `%inc` statement about the location for storing the macro. Most of the above data steps and output formats and destinations will be needed for all of the SAS macros.

Example 2.8.1 assumes that the target values are random (`target=random`); the error structure is constant (`error=const`); the CCC allowance is 0.9775 (`CCC_a=0.9775`); the TDI allowance is 150 mg/dL (`TDI_a=150`); and the CP allowance is 0.9 (`CP_a=0.9`). The combination of TDI allowance of 150 mg/dL and the CP allowance of 0.9 means that we would like to ensure that 90% of the HemoCue observations are within 150 mg/dL of the Sigma values. Note that the CCC allowance does not affect any of the calculations. We may omit it if we do not have an allowance (`CCC_a=.`). We also would like to compute the 95% upper or lower confidence limit for each agreement index (`alpha=0.05`). For creating the agreement plot, we need to specify the data range to be plotted (`min=0`, `max=2000`), with the linear increment of 250 mg/dL (`by=250`) as shown in Fig. 2.1. We would like to format the TDI output with one significant digit (`dec=1`). The `xxx.rtf` output file contains the table and plot as shown in Table 2.1 and Fig. 2.1.

Suppose we would like to assume that the error structure is proportional; we then change from `error=const` to `error=prop`. In this case, we will need to change the plotting increment into a log increment by making `min=25`, `max=3200`, and `by=50 100 200 400 800 1600`. We need to change the TDI allowance to percent change. We do not need to specify the significant digit for the TDI% output, because it is coded to print as a percentage with two significant digits after the decimal point. The code to execute the SAS macro is shown below:

```
%agreement(dataset=e2_1, y=HemoCueAve, x=SigmaAve,
    V_label=HemoCue, H_label=Sigma, min=25, max=3200, by=50 100
    200 400 800 1600, CCC_a=0.9775, CP_a=0.9, TDI_a=50,
    error=prop, target=random, alpha=0.05);
```

We will then have the outputs as shown in Table 7.1 and Fig. 7.1. From the agreement plot of Fig. 7.1, we immediately know that the proportional error assumption is not appropriate, with larger variations at the lower concentrations. Therefore, the results shown in Table 7.1 become irrelevant. When the target values are assumed fixed as in Example 2.8.2, we can simply use `target=fixed`.

Similarly, the corresponding R function `agreement` for Example 2.8.1 can be executed with the following R code:

```
library(Agreement);

HemocueAve=apply(Example2_1[,c("HEMOCUE1", "HEMOCUE2")],1,
                 mean);
SigmaAve=apply(Example2_1[,c("SIGMA1","SIGMA2")],1,mean);

#For constant error structure.

agr_c=agreement(y=HemocueAve,x=SigmaAve,V_label="Hemacue",
H_label="Sigma", min=0, max=2000, by=250, CCC_a=0.9775,
CP_a=0.9,TDI_a=150,error="const", target="random", dec=1,
alpha=0.05)
html.report(agr_c, file="report_1")

#For proportional error structure.

agr_p=agreement(y=HemocueAve,x=SigmaAve,V_label="Hemacue",
H_label="Sigma", min=25, max=3200, by=c(50,100,200,400,800,
1600), CCC_a=0.9775, CP_a=0.9, TDI_a=50, error="prop",
target="random", alpha=0.05)
html.report(agr_p, file="report_2")
```

All the parameters of the R functions have the same definition as the SAS macro, except that there is no dataset parameter, and we must add double quotation marks for some parameters. The function `html.report` is used to generate an html file containing the information as shown in Table 2.1 and Fig. 2.1. For a detailed explanation of the R functions, please read the R help document.

Table 7.1 HemoCue and sigma readings on measuring DCLHb assuming proportional error

Statistics	CCC	Precision	Accuracy	TDI%$_{0.9}$	CP$_{50\%}$	RBS*
Estimate	0.9744	0.9752	0.9992	49.78	0.9001	0.02
95% Conf. limit	0.9691	0.9701	0.9976	54.06	0.8748	.
Allowance	0.9775	.	.	50.00	0.9000	.

$n = 299$.

*The relative bias squared (RBS) must be less than 1 or 8 for CP$_a$ of 0.9 or 0.8, respectively, in order for the approximated TDI to be valid. Otherwise, the TDI estimate is conservative depending on the RBS value.

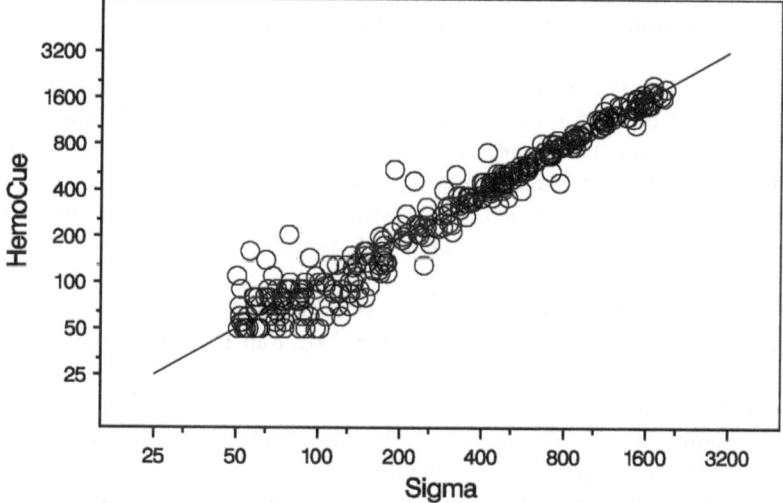

Fig. 7.1 HemoCue and Sigma readings on measuring DCLHb assuming proportional error

7.1.2   *Unified Model*

To assess the intraassay, interassay, and total-assay agreement as shown in Example 5.9.1, we need to call the macro using the %UnifiedAgreement statement with parameters defined below:

- Dataset: name of your data. We must avoid using dataset names a, one, y, x, id, rating, outb, outc, method, best, par, r, and r1, because these dataset names are used in this macro.
- k: number of methods/raters/instruments/assay, etc.
- m: number of replications for each k.
- var: dependent variable names to be evaluated, e.g., y1_1, y1_2, ..., y1_m, y2_1, y2_2, ..., y2_m, ..., yk_1, yk_2, ..., yk_m, etc.
- CCC_a: CCC allowance when m = 1.
- CCC_a_intra: intra-CCC allowance.

- `CCC_a_inter`: inter-CCC allowance.
- `CCC_a_total`: total CCC allowance.
  The above CCC allowances can be set as missing if there are no prespecified allowances.
- `CP_a`: coverage probability (CP) allowance for continuous data.
- `TDI_a`: TDI allowance when $m = 1$ for continuous data.
- `TDI_a_intra`: intra-TDI allowance for continuous data.
- `TDI_a_inter`: inter-TDI allowance for continuous data.
- `TDI_a_total`: total TDI allowance for continuous data.
  The above `TDI_a` values must be specified as a percent change when `Error=prop` is specified.
- `tran = 1`: transformation such as Z, logit, and log will be used for statistical inference.
  `tran = 0`: no transformation will be used for statistical inference.
  `tran=1` can be used for categorical data, but the TDI and CP outputs would become irrelevant. Therefore, `tran=0` is recommended for all categorical data and `tran=1` is recommended for all continuous data.
- `Error = const`: constant error structure for continuous data. When `error=const`, TDI is an absolute difference with the same measurement unit as for the original data.
  `Error = prop`: proportional error structure for continuous data. When `error=prop`, TDI is a percent change. Log transformation will be applied to original data.
  For categorical data, use `Error = const`.
- `Dec`: significant digits after the decimal point printed for TDI when `error=const` is specified. The default is `dec=2`.
- `Alpha`: $(1 - \alpha)\%$ one-tailed confidence limit. The default is 0.05.

For calculating intraagreement, interagreement, and total-agreement indices as shown in Example 5.9.1, we would execute the following code. Note that `UnifiedAgreement.sas` does not produce agreement plots. Therefore, the code of the first three `%agreement` macros shown below is meant to produce the plots of within HemoCue, within Sigma, and between the two methods based on the average.

```
%agreement(dataset=ex.example2_1, y=hemocue2,x=hemocue1,
  min=0, max=2000, by=250, V_label=HemoCue 2, H_label=HemoCue 1,
error=const, target=random, CCC_a=0.9775, CP_a=0.9, alpha=0.05,
TDI_a=150, dec=1);

%agreement(dataset=ex.example2_1, y=Sigma2, x=sigma1,
  min=0, max=2000, by=250, V_label=Sigma 2, H_label=Sigma 1,
error=const, target=random, CCC_a=0.9775, CP_a=0.9, alpha=0.05,
TDI_a=150, dec=1);

%agreement(dataset=e2_1, y=HemocueAve, x=SigmaAve,
  min=0,max=2000, by=250, V_label=HemoCue, H_label=Sigma,
error=const,target=random, CCC_a=0.9775, CP_a=0.9, alpha=0.05,
TDI_a=150, dec=1);
```

```
%UnifiedAgreement (dataset=ex.example2_1,var=hemocue1
   hemocue2 sigma1 sigma2,k=2,m=2,CCC_a_intra=0.9943,
   CCC_a_inter=0.9775,CCC_a_total=0.9775,
CP_a=0.9, tran=1, TDI_a_intra=75, TDI_a_inter=150,
TDI_a_total=150, error=const, dec=1, alpha=0.05);
```

In UnifiedAgreement.sas for Example 5.9.1, the parameter for specifying dependent variables must be in the order y1_1, y1_2, ..., y1_m, y2_1, y2_2, ..., y2_m, ..., and yk_1, yk_2, ..., yk_m (var=hemocue1 hemocue2 sigma1 sigma2). We need to specify the number of raters (k=2) and the number of replicates (m=2) per rater per subject. We specify the CCC allowances for intra (CCC_a_intra=0.9943), inter (CCC_a_inter=0.9775), and total (CCC_a_total=0.9775), and the TDI allowances for intra (TDI_a_intra=75), inter (TDI_a_inter=150), and total (TDI_a_total=150). The CP allowance is set at 0.9 to capture 90% of observations (CP_a=0.9) for all intra, inter, and total CPs. We specify tran=1, indicating that Z-transformation for CCCs, logit transformation for CPs and accuracy coefficients, and log transformation for TDIs will be used for statistical inferences. For categorical data, we may specify tran=0, since the above transformations are not necessarily needed. All TDI and CP indices would become irrelevant for categorical data and thus will not be printed. The results are shown in Example 5.9.1.

Similarly, the corresponding R function unified.agreement for Example 5.9.1 can be performed with the following code:

```
ua=unified.agreement (dataset=Example2_1,
   var=c("HEMOCUE1","HEMOCUE2", "SIGMA1","SIGMA2"),
   k=2, m=2, CCC_a_intra=0.9943,
CCC_a_inter=0.9775, CCC_a_total=0.9775, CP_a=0.9, tran=1,
TDI_a_intra=75, TDI_a_inter=150, TDI_a_total=150,
error="const", dec=1, alpha=0.05) html.unified_agreement (ua)
```

The parameter dataset is the name of the dataset, which must contain the variables in the var parameter. The order of the entries specified in the parameter var should follow the same rule of that in the SAS macro. If the parameter var is not given, the R function will use all the variables in the input dataset. The other parameters have the same definition as the SAS macro. The function html.unified_agreement is used to generate an html file containing the summary table of unified agreement.

7.1.2.1   Unified Model with m=1

The macro UnifiedAgreement.sas can also be applied to cases in which there is no replicates (m=1). When $k = 2$, the estimated confidence limit is slightly different from that estimated by the macro agreement.sas, because the latter

Table 7.2 HemoCue and Sigma readings on measuring DCLHb using GEE

Statistics	CCC	Precision	Accuracy	TDI%$_{0.9}$	CP$_{150}$	RBS*
Estimate	0.9866	0.9866	1.0000	127.3	0.9474	0.00
95% Conf. limit	0.9825	0.9825	0.9987	145.9	0.9228	.
Allowance	0.9775	.	.	150.0	0.9000	.

For k = 2, n = 299, and m = 1.
*The relative bias squared (RBS) must be less than 1 or 8 for CP$_a$ of 0.9 or 0.8, respectively, in order for the approximated TDI to be valid.

assumes normality for deriving the variances of the estimates of the agreement indices, while the former uses the GEE approach without assuming normality. For more robustness of the confidence limits and/or when k > 2, we might need to use the unified macro. However, to produce the agreement plots, we would need to call the `agreement.sas` macro. In addition, the definitions of precision and accuracy are slightly different because the unified approach assumes that the variances of assays or raters are equal, and it utilizes the approximation according to (5.62) for CP.

For Example 2.8.1, if we want to use the more robust GEE approach for the case of m = 1, we can run the following code after taking the average of the replicates for HemoCue and Sigma as we did with running %agreement earlier:

```
%UnifiedAgreement (dataset=e2_1, var=HemocueAve SigmaAve,
  k=2, m=1, CCC_a=0.9775,CP_a=0.9, tran=1, TDI_a=150,
  error=const, dec=1,alpha=0.05);
```

The results are shown in Table 7.2. As expected, these are exactly the same as the results shown in Table 5.1 under interagreement. Compared to Table 2.1, the CCC and TDI estimates are identical, and the precision and accuracy coefficients are almost the same. The CP estimate using GEE is 0.947 compared to 0.946 in Table 2.1. These are almost identical because of an almost perfect accuracy, indicating that their variances are the same. The lower confidence limits for CCC, precision and accuracy coefficients, and CP are slightly smaller than those shown in Table 2.1. Correspondingly, the upper confidence limit for TDI is slightly larger than those shown in Table 2.1, indicating that for this example the GEE approach is slightly more conservative.

Similarly, the corresponding R function for Example 2.8.1 by the unified approach can be performed with the following code:

```
unified.agreement (dataset=cbind(HemocueAve,SigmaAve),
  k=2,m=1,CCC_a=0.9775,CP_a=0.9, tran=1, TDI_a=150,
  error="const",dec=1,alpha=0.05);
```

7.1.3 TIR and IIR

To calculate the TIR and IIR indices introduced in Chapter 6, we need to call the SAS macro `TIR_IIR.sas` using the `%TIR_IIR` statement with parameters defined below:

- `dataset`: name of your data. We must avoid using dataset names a, b, c, t1, t2, ttt, bt, one, final, because these dataset names are used in this macro.
- `k`: number of methods/raters/instruments/assay, etc.
- `m`: number of replications for each `k`.
- `var`: dependent variable names to be evaluated, e.g., y1_1, y1_2, ..., y1_m, y2_1, y2_2, ..., y2_m, ..., yk_1, yk_2, ..., yk_m, etc.
- `TIR_test`: the selected test raters for calculating TIR; must be input in the format (`'1'`, `'2'`, `'3'`, ... `'k'`), where `'1'` represents the first m columns for rater 1, `'2'` represents the second m columns for rater 2, and `'k'` represents the last m columns for rater k, etc. When calculating multiple TIRs, the test raters for calculating each TIR must be separated by #. For example, when k = 3, we specify (`'1'`,`'3'`)#(`'1'`,`'2'`,`'3'`)#(`'3'`)#(`'2'`)#(`'1'`,`'2'`) for each of the five sets of the test raters.
- `TIR_ref`: the selected reference raters for calculating TIR that correspond to `TIR_test`. If `ref=(all)` is specified, then the intraraters of all raters will be used as the denominator. When calculating multiple TIRs, the corresponding reference raters must be separated by #. For example, use (`'2'`)#(all)#(`'1'`,`'2'`)#(`'1'`)#(`'1'`,`'2'`) to represent the five selected sets of reference raters. When `TIR_ref` is not specified as (all), each TIR is computed as the total MSD of test vs selected reference raters relative to the intra MSD of the selected reference raters. When `TIR_ref` is specified as (all), the macro will assess the average of the total MSD of all raters relative to the average of intra MSD of all raters. For the first TIR example as shown in `TIR_test` and `TIR_ref`, the macro would assess the average of the total MSD of "raters 1 vs 2 and raters 3 vs 2" relative to the intra MSD of "rater 2." For the second TIR example, the macro would assess the average of the total MSD of all raters relative to the average of intra MSD of all raters. For the third TIR example, the macro would assess the average of the total MSD of "raters 3 vs 1 and raters 3 vs 2" relative to the average of intra MSD of "raters 1 and 2." For the fourth TIR example, the macro would assess the total MSD of "raters 2 and 1" relative to the intra MSD of "rater 1." For the fifth TIR example, the macro would assess the total MSD of "raters 1 and 2" relative to the average of intra MSD of "raters 1 and 2."
- `IIR_test`: the selected test raters for calculating IIR, which must be input in the format of (`'1'`,`'2'`,`'3'`, ...`'k'`). When calculating multiple IIRs, the test raters for calculating each IIR must be separated by #. For example, when k = 3, specify (`'1'`)#(`'2'`)#(`'3'`)#(`'1'`).

Table 7.3 TIR and IIR between HemoCue ('1') and Sigma ('2') with duplicates

	TIR	IIR
Statistics	Total$_{('1','2')vs(all)}$/Intra$_{(all)}$	Intra$_{('1')}$/Intra$_{('2')}$
Estimate	10.094	0.348
95% Conf. limit*	13.657	(0.189, 0.643)
Compared to	.	1.00

For k = 2, n = 299, and m = 2.
*One-tailed upper limit for TIR and two-tailed interval for IIR.

- `IIR_ref`: the selected reference raters for calculating IIR. When calculating multiple IIRs, the corresponding reference raters must be separated by #. For example, specify `('2','3')#('1','3')#('1','2')#('2')`. Each set of reference raters must be mutually exclusive from its corresponding set of the selected test raters.
- `Error = const` for the constant error structure for continuous data.
  `Error = prop` for the proportional error structure for continuous data. Here, log transformation to data will be applied for continuous data. For categorical data, use `Error=const`.
- Alpha: $100(1 - \alpha)\%$ one-tailed upper confidence limit for TIR or two-tailed confidence interval for IIR. The default is 0.05.
- `TIR_a`: allowance for TIR.

To calculate TIR and IIR for Example 5.9.1, we would execute the following code:

```
%TIR_IIR(dataset=ex.example2_1, var=hemocue1 hemocue2
  sigma1 sigma2, k=2, m=2, TIR_test=('1','2'), TIR_ref=(all),
  IIR_test=('1'),IIR_ref=('2'),
error=const, alpha=0.05, TIR_a=.);
```

The results are shown in Table 7.3. The TIR of MSD$_{total}$ relative to MSD$_{intra}$ was estimated to be 10.09 with one-sided 95% upper confidence limit of 13.66, which was much larger than any clinically meaningful criterion, as evident by comparing Fig. 5.3 to Figs. 5.1 and 5.2. The IIR of MSD$_{intra\_HemoCue}$ relative to MSD$_{intra\_Sigma}$ was estimated to be 0.348 with the 95% confidence interval of 0.189–0.643, which indicates that HemoCue had better within-assay precision than Sigma, as evident by comparing Figs. 5.1 and 5.2.

Similarly, the corresponding R functions `TIR_IIR` for the comparative agreement approach can be performed with the following code:

```
TIR_IIR(dataset=Example2_1,var=c("HEMOCUE1", "HEMOCUE2",
"SIGMA1", "SIGMA2"),
k=2,m=2,TIR_test=c("1,2"),TIR_ref=c("All"),IIR_test=c("1"),
IIR_ref=c("2"), error="const", alpha=0.05, TIR_a=.);
```

Table 7.4 Agreement statistics among three examiners based on
reading 1

Statistics	CCC	Precision	Accuracy
Estimate	0.4958	0.5034	0.9849
95% Conf. limit	0.4289	0.4370	0.9745
Allowance	.	.	.

For k = 3, n = 400, and m = 1.

All the parameters have the same definition as in the SAS macro. However, the format of the parameters TIR_test, TIR_ref, IIR_test, and IIR_ref are slightly different: TIR_test=c("1,2") means that the selected test raters for calculating TIR are the first and second raters. If there are multiple TIRs, each set of test raters must be an entry in the sequence using double quotation marks separated by a comma. For example, when k = 3, we may specify TIR_test=c("1,3","1,2,3","3","2","1,2"). The formats of TIR_ref, IIR_test, and IIR_ref are defined similarly.

7.2   Workshop for Categorical Data

7.2.1   The Basic Model (m=1)

We begin with the workshop for categorical data using Example 5.9.3, and examine the kappa of three examiners based on their first readings. These frequency tables can be seen in Tables 5.10 and 5.10. In this example, the variable names for the three examiners and their duplicates are m1_1, m1_2, m2_1, m2_2, m3_1, m3_2. We are now interested in the kappa only of m1_1, m2_1, m3_1. We then execute the following code:

```
%UnifiedAgreement(dataset=ex.Example5_3, var=m1_1 m2_1
   m3_1, k=3, m=1, ccc_a=., tran=0, alpha=0.05);
```

The results are shown in Table 7.4. These are slightly lower than those shown under total agreement in Table 5.13. Again, the results show that the disagreement was largely due to imprecision rather than inaccuracy. For categorical data, TDI and CP are not meaningful, and therefore are not computed.

Similarly, the corresponding R macro for Example 5.9.3 using the unified approach can be performed with the following code:

```
unified.agreement(dataset=Example5_3, var=c("m1_1",
   "m2_1","m3_1"), k=3, m=1, CCC_a=NA, tran=0, alpha=0.05);
```

Table 7.5 Agreement statistics among the first two examiners based on reading 1

Statistics	CCC	Precision	Accuracy
Estimate	0.5147	0.5148	0.9998
95% Conf. limit	0.4225	0.4226	0.9982
Allowance	.	.	.

For k = 2, n = 400, and m = 1.

7.2.1.1 Equivalence to SAS Procedure FREQ when k=2 and m=1

When $k = 2$, we can also compute the kappa-related results by running the SAS procedure FREQ. To demonstrate the equivalence, we first examine the agreement between the first readings of examiners 1 and 2 by executing the following code:

```
%UnifiedAgreement(dataset= ex.Example5_3, var=m1_1 m2_1,
   k=2, m=1, CCC_a=., tran=0, alpha=0.05);
```

The results are displayed in Table 7.5.
We then execute the following SAS code:

```
proc freq data=ex.Example5_3;
table m1_1*m2_1/agree (WT=FC) alpha=0.1;
run;
```

Note that we use an alpha value of 0.1 because we want only the one-tailed lower confidence limit. The (WT=FC) is not necessary in this case because this example has a binary outcome, but we leave it there just in case the data have an ordinal outcome and we would like to use the square distance function. The results are shown in the following SAS output file:

Simple Kappa Coefficient	
Kappa	0.5147
ASE	0.0560
90% Lower Conf. limit	0.4225
90% Upper Conf. limit	0.6069
Sample Size = 400	

The kappa and its lower confidence limit are exactly the same as shown in Table 7.4. The proof of such equivalence is shown in Section 5.7.2. Note that this SAS procedure FREQ cannot perform kappa for more than two raters.

7.2.2 Unified Model

To calculate the intraassay, interassay, and total-assay agreement indices for categorical data with $k \geq 2$ and $m \geq 1$ using the data of Example 5.9.3, we execute the

Table 7.6 TIR and IIR among three examiners with duplicates

Statistics	TIR $\text{Total}_{('1','2','3')vs(all)}/\text{Intra}_{(all)}$	IIR $\text{Intra}_{('3')}/\text{Intra}_{('1','2')}$
Estimate	1.560	1.323
95% Conf. limit*	2.376	(0.931, 1.881)
Compared to	2.5	1.00

For k = 3, n = 400, and m = 2.
*One-tailed upper limit for TIR and two-tailed interval for IIR.

following code. The results are shown in Table 5.13 of Chapter 5. The frequency tables can be seen in Tables 5.7–5.12.

```
%UnifiedAgreement(dataset=ex.Example5_3, var=m1_1 m1_2
  m2_1 m2_2 m3_1 m3_2, k=3,m=2,CCC_a_intra=0.64,
  CCC_a_inter=0.51,CCC_a_total=.,tran=0, alpha=0.05);
```

Similarly, the corresponding R macro for Example 5.9.3 with the unified approach can be performed with the following code:

```
unified.agreement(dataset=Example5_3,k=3,m=2,CCC_a_intra=0.64,
CCC_a_inter=0.51, CCC_a_total=NA, tran=0, alpha=0.05);
```

7.2.3 TIR and IIR

To study the TIR and IIR information as shown in Example 6.7.2 using the same data in Example 5.9.3, we execute the following code:

```
%TIR_IIR(dataset=data5.Example5_3, var=m1_1-m1_2
  m2_1-m2_2 m3_1-m3_2, k=3,m=2, TIR_test=('1','2','3'),
  TIR_ref=(all), IIR_test=('3'), IIR_ref=('1','2'),
  error=const, alpha=0.05, TIR_a=2.5)
```

The results are shown in Table 7.6, with the description for TIR given in Section 6.7.2.

Similarly, the corresponding R macro for the comparative approach can be performed with the following code:

```
TIR_IIR(dataset=Example5_3,var=c("m1_1","m1_2","m2_1",
  "m2_2","m3_1","m3_2"), k=3, m=2,TIR_test=c("1,2,3"),
  TIR_ref=c("All"), IIR_test=c("3"), IIR_ref=c("1,2"),
  error="const", alpha=0.05, TIR_a=2.5);
```

Table 7.7   TIR and IIR between test (1) and reference (2) compounds
with duplicates

Statistics	TIR Total$_{('1')vs('2')}$/Intra$_{('2')}$	IIR Intra$_{('1')}$/Intra$_{('2')}$
Estimate	0.6907	0.4324
95% Conf. limit*	1.0761	(0.1676, 1.1151)
Compared to	2.25	1.00

*For k = 2, n = 39, and m = 2.
One-tailed upper limit for TIR and two-tailed interval for IIR.

7.3   Individual Bioequivalence

In Example 6.7.4, we first need to compute TDI%$_{0.8}$ between the duplicate values
of the reference compound, namely, R1 and R2. We can either use the basic macro
`agreement.SAS` or the unified model `UnifiedAgreement.sas`. For using
the basic model, we execute the following code:

```
%agreement(dataset=ex.Example6_4, y=R1, x=R2,
  V_label=R1, H_label=R2, min=2, max=64, by=4 8 16, CCC_a=0.9,
  CP_a=0.8, TDI_a=50, error=prop, target=random, alpha=0.05);
```

The TDI%$_{0.8}$ was estimated to be 124.4%, which is greater than 43.7%,
indicating that the reference scale should be used. For calculating the TIR and IIR
with their 95% confidence limits, we execute the following code:

```
%TIR_IIR(dataset=ex.Example6_4, var=t1 t2 r1 r2, k=2,
  m=2, TIR_test=('1'), TIR_ref=('2'), IIR_test=('1'),
  IIR_ref=('2'), error=prop, alpha=0.05, TIR_a=2.25);
```

The results are shown in Table 7.7, with the description for TIR given in the last
paragraph of Section 6.7.2.

Similarly, the corresponding R macro for the comparative approach can be
performed with the following code:

```
agreement(y=Example6_4[,"R1"], x=Example6_4[,"R2"],
  V_label="R1", H_label="R2", min=2, max=64, by=c(4,8,16),
  CCC_a=0.9, CP_a=0.8, TDI_a=50, error="prop",
  target="random", alpha=0.05);

TIR_IIR(dataset=Example6_4, var=c("T1","T2","R1","R2"),
  k=2, m=2, TIR_test=c("1"), TIR_ref=c("2"), IIR_test=c("1"),
  IIR_ref=c("2"), error="prop", alpha=0.05, TIR_a=2.25);
```

References

Agresti, A. 1990. *Categorical data analysis*. New York: Wiley.

Agresti, A. 1996. *An Introduction to Categorical Data Analysis*. New York: Wiley.

Aickin, M. 1983. *Linear Statistical Analysis of Discrete Data*. New York: Wiley.

Barnhart, H., M. Haber, and L. Lin. 2007. An overview on assessing agreement with continuous measurements. *Journal of Biopharmaceutical Statistics* 17(4):529–569.

Barnhart, H., M. Haber, and J. Song. 2002. Overall concordance correlation coefficient for evaluating agreement among multiple observers. *Biometrics* 58(4):1020–1027.

Barnhart, H., A. Kosinski, and M. Haber. 2007. Assessing individual agreement. *Journal of Biopharmaceutical Statistics* 17(4):697–719.

Barnhart, H., J. Song, and M. Haber. 2005. Assessing intra, inter and total agreement with replicated readings. *Statistics in Medicine* 24(9):1371–1384.

Barnhart, H., J. Song, and R. Lyles. 2005. Assay validation for left-censored data. *Statistics in Medicine* 24(21):3347–3360.

Barnhart, H. and J. Williamson. 2001. Modeling concordance correlation via GEE to evaluate reproducibility. *Biometrics* 57(3):931–940.

Bartko, J. 1966. The intraclass correlation coefficient as a measure of reliability. *Psychological Reports* 19(1):3–11.

Birch, M. 1964. The detection of partial association, I: the 2 x 2 case. *Journal of the Royal Statistical Society, Series B* 26(3):313–324.

Birch, M. 1965. The detection of partial association, II: the general case. *Journal of the Royal Statistical Society. Series B* 27(1):111–124.

Bland, J. and D. Altman. 1999. Measuring agreement in method comparison studies. *Statistical Methods in Medical Research* 8(2):135–160.

Bland, J. and D.G. Altman. 1986. Statistical methods for assessing agreement between two methods of clinical measurement. *The LANCET* i:307–310.

Bloch, D. and H. Kraemer. 1989. 2× 2 kappa coefficients: measures of agreement or association. *Biometrics* 45(1):269–287.

Brennan, R. 2001. *Generalizability Theory*. New York: Springer.

Broemeling, L. 2009. *Bayesian Methods for Measures of Agreement*. Boca Raton: Chapman & Hall/CRC.

Bross, I. 1985. Why proof of safety is much more difficult than proof of hazard. *Biometrics* 41(3): 785–793.

Carrasco, J. and L. Jover. 2003. Estimating the generalized concordance correlation coefficient through variance components. *Biometrics* 59(4):849–858.

Carrasco, J. and L. Jover. 2005. Concordance correlation coefficient applied to discrete data. *Statistics in Medicine* 24(24):4021–4034.

L. Lin et al., *Statistical Tools for Measuring Agreement*,
DOI 10.1007/978-1-4614-0562-7, © Springer Science+Business Media, LLC 2012

Carrasco, J., T. King, and V. Chinchilli. 2009. The concordance correlation coefficient for repeated measures estimated by variance components. *Journal of Biopharmaceutical Statistics* 19(1):90–105.

Carrasco L., J. Luis, T. King, and V. Chinchilli. 2007. Comparison of concordance correlation coefficient estimating approaches with skewed data. *Journal of Biopharmaceutical Statistics* 17: 673–684.

Chen, C. and H. Barnhart. 2008. Comparison of ICC and CCC for assessing agreement for data without and with replications. *Computational Statistics & Data Analysis* 53(2):554–564.

Chinchilli, V., J. Martel, S. Kumanyika, and T. Lloyd. 1996. A weighted concordance correlation coefficient for repeated measurement designs. *Biometrics* 52(1):341–353.

Choudhary, P. 2007. A tolerance interval approach for assessment of agreement with left censored data. *Journal of Biopharmaceutical Statistics* 17(4):583–594.

Choudhary, P. 2008. A tolerance interval approach for assessment of agreement in method comparison studies with repeated measurements. *Journal of Statistical Planning and Inference* 138(4):1102–1115.

Choudhary, P. and H. Nagaraja. 2007. Tests for assessment of agreement using probability criteria. *Journal of Statistical Planning and Inference* 137(1):279–290.

Christensen, R. 1997. *Log-linear Models and Logistic Regression* 2nd Ed. New York: Springer.

Cicchetti, D. and T. Allison. 1971. A new procedure for assessing reliability of scoring EEG sleep recordings. *American Journal of EEG Technology* 11:101–109.

Cicchetti, D. and J. Fleiss. 1977. Comparison of the null distributions of weighted kappa and the C ordinal statistic. *Applied Psychological Measurement* 1(2):195–201.

CLIA Final Rule 2003. CLIA programs; laboratory requirements relating to quality systems and certain personnel qualifications. Final rule. *Federal Register* 68(16):3639–3714. available at http://www.phppo.cdc.gov/clia/pdf/CMS-2226-F.pdf.

Cochran, W. 1950. The comparison of percentages in matched samples. *Biometrika* 37(3-4): 256–266.

Cohen, J. 1960. A coefficient of agreement for nominal scales. *Educational and Psychological Measurement* 20(1):37–46.

Cohen, J. 1968. Weighted kappa: Nominal scale agreement provision for scaled disagreement or partial credit. *Psychological Bulletin* 70(4):213–220.

Cox, D. and E. Snell. 1989. *Analysis of Binary Data*. Boca Raton: Chapman & Hall/CRC.

Darroch, J. 1981. The Mantel-Haenszel test and tests of marginal symmetry; fixed-effects and mixed models for a categorical response. *International Statistical Review* 49:285–307.

Davis, C. and D. Quade. 1968. On comparing the correlations within two pairs of variables. *Biometrics* 24(4):987–995.

Donner, A. and M. Eliasziw. 1992. A goodness-of-fit approach to inference procedures for the kappa statistic: Confidence interval construction, significance-testing and sample size estimation. *Statistics in Medicine* 11(11):1511–1519.

Dunnett, C. and M. Gent. 1977. Significance testing to establish equivalence between treatments, with special reference to data in the form of 2 x 2 tables. *Biometrics* 33(4):593–602.

Escaramis, G., C. Ascaso, and J. Carrasco. 2010. The total deviation index estimated by tolerance intervals to evaluate the concordance of measurement devices. *BMC Medical Research Methodology* 10(1):31.

Everitt, B. 1968. Moments of the statistics kappa and weighted kappa. *British Journal of Mathematical and Statistical Psychology* 21(1):97–103.

Fisher, S. 1925. *Statistical Methods for Research Workers*. Edinburgh: Oliver & Boyd.

Fleiss, J. 1971. Measuring nominal scale agreement among many raters. *Psychological Bulletin* 76(5):378–382.

Fleiss, J.L. 1973. *Statistical Methods for Rates and Proportions*. New York: John Wiley & Sons.

Fleiss, J. 1986. Reliability of measurement. *The Design and Analysis of Clinical Experiments* 1(1):1–32.

Fleiss, J. and J. Cohen. 1973. The equivalence of weighted kappa and the intraclass correlation coefficient as measures of reliability. *Educational and Psychological Measurement* 33(3): 613–619.

Fleiss, J., J. Cohen, and B. Everitt. 1969. Large sample standard errors of kappa and weighted kappa. *Psychological Bulletin* 72(5):323–327.

Fleiss, J. and J. Cuzick. 1979. The reliability of dichotomous judgments: Unequal numbers of judges per subject. *Applied Psychological Measurement* 3(4):537–542.

Fleiss, J. and B. Everitt. 1971. Comparing the marginal totals of square contingency tables. *British Journal of Mathematical and Statistical Psychology* 24:117–123.

Fleiss, J., B. Levin, M. Paik, and J. Wiley. 1981. *Statistical Methods for Rates and Proportions* 2nd Ed. New York: Wiley.

Fleiss, J. and P. Shrout. 1978. Approximate interval estimation for a certain intraclass correlation coefficient. *Psychometrika* 43(2):259–262.

Freeman, D. 1987. *Applied Categorical Data Analysis*. New York: Marcel Dekker.

Friedman, M. 1937. The use of ranks to avoid the assumption of normality implicit in the analysis of variance. *Journal of the American Statistical Association* 32(200):675–701.

Goodman, L. 1978. *Analyzing Qualitative/Categorical Data: Log-linear Models and Latent-Structure Analysis*. Cambridge, MA: Abt Books.

Grant, E. and R. Leavenworth. 1972. *Statistical Quality Control*. New York: McGraw-Hill.

Guo, Y. and A. Manatunga. 2007. Nonparametric estimation of the concordance correlation coefficient under univariate censoring. *Biometrics* 63(1):164–172.

Guo, Y. and A. Manatunga. 2009. Measuring agreement of multivariate discrete survival times using a modified weighted kappa coefficient. *Biometrics* 65(1):125–134.

Haber, M. and H. Barnhart. 2008. A general approach to evaluating agreement between two observers or methods of measurement from quantitative data with replicated measurements. *Statistical Methods in Medical Research* 17(2):151–169.

Haber, M., J. Gao, and H. Barnhart. 2007. Assessing observer agreement in studies involving replicated binary observations. *Journal of Biopharmaceutical Statistics* 17(4):757–766.

Haberman, S. 1974. *The Analysis of Frequency Data*. Chicago: University of Chicago Press.

Haberman, S. 1978. *Analysis of Qualitative Data, volumn 1. Introductory Topics*. New York: Academic Press.

Haberman, S. 1979. *Analysis of Qualitative Data, volumn 2. New Developments*. New York: Academic Press.

Hedayat, A., C. Lou, and B. Sinha. 2009. A Statistical Approach to Assessment of Agreement Involving Multiple Raters. *Communications in Statistics-Theory and Methods* 38(16): 2899–2922.

Helenowski, I., E. Vonesh, H. Demirtas, A. Rademaker, V. Ananthanarayanan, P. Gann, and B. Jovanovic. 2011. Defining Reproducibility Statistics as a Function of the Spatial Covariance Structures in Biomarker Studies. *The International Journal of Biostatistics* 7(1):article2.

Hiriote, S. and V. Chinchilli. 2010. Matrix-based concordance correlation coefficient for repeated measures. *Biometrics* 66:1–20.

Holder, D. and F. Hsuan. 1993. Moment-based criteria for determining bioequivalence. *Biometrika* 80(4):835–846.

King, T. and V. Chinchilli. 2001a. A generalized concordance correlation coefficient for continuous and categorical data. *Statistics in Medicine* 20(14):2131–2147.

King, T. and V. Chinchilli. 2001b. Robust estimators of the concordance correlation coefficient. *Journal of Biopharmaceutical Statistics* 11(3):83–105.

King, T., V. Chinchilli, and J. Carrasco. 2007. A repeated measures concordance correlation coefficient. *Statistics in Medicine* 26(16):3095–3113.

King, T., V. Chinchilli, K. Wang, and J. Carrasco. 2007. A class of repeated measures concordance correlation coefficients. *Journal of Biopharmaceutical Statistics* 17(4):653–672.

Koch, G., J. Landis, J. Freeman, D. Freeman Jr, and R. Lehnen. 1977. A general methodology for the analysis of experiments with repeated measurement of categorical data. *Biometrics* 33(1): 133–158.

Landis, J. and G. Koch. 1977a. A one-way components of variance model for categorical data. *Biometrics* 33(4):671–679.

Landis, J. and G. Koch. 1977b. An application of hierarchical kappa-type statistics in the assessment of majority agreement among multiple observers. *Biometrics* 33(2):363–374.

Landis, J. and G. Koch. 1977c. The measurement of observer agreement for categorical data. *Biometrics* 33(1):159–174.

Landis, J., T. Sharp, S. Kuritz, and G. Koch. 1998. Mantel–haenszel methods. *Encyclopedia of Biostatistics* 3:2378–2391.

Li, R. and M. Chow. 2005. Evaluation of reproducibility for paired functional data. *Journal of Multivariate Analysis* 93(1):81–101.

Liang, K. and S. Zeger. 1986. Longitudinal data analysis using generalized linear models. *Biometrika* 73(1):13–22.

Lin, L. 1989. A concordance correlation coefficient to evaluate reproducibility. *Biometrics* 45(1): 255–268.

Lin, L. 1992. Assay validation using the concordance correlation coefficient. *Biometrics* 48(2): 599–604.

Lin, L. 2000. Total deviation index for measuring individual agreement with applications in laboratory performance and bioequivalence. *Statistics in Medicine* 19(2):255–270.

Lin, L. 2003. Measuring agreement. *Encyclopedia of Biopharmaceutical Statistics*, 561–567.

Lin, L. 2008. Overview of agreement statistics for medical devices. *Journal of Biopharmaceutical Statistics* 18(1):126–144.

Lin, L. and V. Chinchilli. 1997. Rejoinder to the letter to the editor from Atkinson and Nevill. *Biometrics* 53(2):777–778.

Lin, L., A. Hedayat, B. Sinha, and M. Yang. 2002. Statistical Methods in Assessing Agreement. *Journal of the American Statistical Association* 97(457):257–270.

Lin, L., A. Hedayat, and Y. Tang. 2012. A comparison for measuring individual agreement. *Journal of Biopharmaceutical Statistics*, accept for publication.

Lin, L., A. Hedayat, and W. Wu. 2007. A unified approach for assessing agreement for continuous and categorical data. *Journal of Biopharmaceutical Statistics* 17(4):629–652.

Lin, L. and L. Torbeck. 1998. Coefficient of accuracy and concordance correlation coefficient: new statistics for methods comparison. *PDA Journal of Pharmaceutical Science and Technology* 52(2):55–59.

Lin, L. and E. Vonesh. 1989. An empirical nonlinear data-fitting approach for transforming data to normality. *American Statistician* 43(4):237–243.

Linn, S. 2004. A new conceptual approach to teaching the interpretation of clinical tests. *Journal of Statistical Education* 12(3):1–9.

Liu, X., Y. Du, J. Teresi, and D. Hasin. 2005. Concordance correlation in the measurements of time to event. *Statistics in Medicine* 24(9):1409–1420.

Lou, C. 2006. *Assessment of Agreement, PhD Thesis*. University of Illinois at Chicago.

Madansky, A. 1963. Tests of homogeneity for correlated samples. *Journal of the American Statistical Association* 58(301):97–119.

McNemar, Q. 1947. Note on the sampling error of the difference between correlated proportions or percentages. *Psychometrika* 12(2):153–157.***

Quiroz, J. 2005. Assessment of equivalence using a concordance correlation coefficient in a repeated measurements design. *Journal of Biopharmaceutical Statistics* 15(6):913–928.

Quiroz, J. and R. Burdick. 2009. Assessment of Individual Agreements with Repeated Measurements Based on Generalized Confidence Intervals. *Journal of Biopharmaceutical Statistics* 19(2):345–359.

Robieson, W. 1999. *On Weighted Kappa and Concordance Correlation Coefficinet, PhD Thesis*. Chicago: University of Illinois.

Rodary, C., C. Com-Nougue, and M. Tournade. 1989. How to establish equivalence between treatments: A one-sided clinical trial in paediatric oncology. *Statistics in Medicine* 8(5): 593–598.

Schuster, C. 2001. Kappa as a parameter of a symmetry model for rater agreement. *Journal of Educational and Behavioral Statistics* 26(3):331–342.

Serfling, R. 1980. *Approximation Theorems of Mathematical Statistics*. New York: John Wiley & Sons.

Sheiner, L. 1992. Bioequivalence revisited. *Statistics in Medicine* 11(13):1777–1788.

Shoukri, M. M. 2004. *Measures of Interobserver Agreement*. Chapman and Hall.

Shoukri, M. and V. Edge. 1996. *Statistical Methods for Health Sciences*. New York: CRC Press.

Shoukri, M. and S. Martin. 1995. Maximum likelihood estimation of the kappa coefficient from models of matched binary responses. *Statistics in Medicine* 14(1):83–99.

Shrout, P. and J. Fleiss. 1979. Intraclass correlations: uses in assessing rater reliability. *Psychol Bull* 86(2):420–428.

Tang, Y. 2010. *A Comparison Model for Measuring Individual Agreement, PhD Thesis*. Chicago: University of Illinois.

Von Eye, A. and E. Mun. 2005. *Analyzing Rater Agreement: Manifest Variable Methods*. New Jersey: Lawrence Erlbaum.

von Eye, A. and C. Schuster. 2000. Log-linear model for rater agreement. *Multiciencia* 4:38–56.

Vonesh, E. and M. Chinchilli. 1997. *Linear and Nonlinear Models for the Analysis of Repeated Measurements*. New York: Marcel Dekker, Inc.

Vonesh, E., V. Chinchilli, and K. Pu. 1996. Goodness-of-fit in generalized nonlinear mixed-effects models. *Biometrics* 52(2):572–587.

Wang, W. and J. Gene Hwang. 2001. A nearly unbiased test for individual bioequivalence problems using probability criteria. *Journal of Statistical Planning and Inference* 99(1):41–58.

Westlake, W. 1976. Symmetrical confidence intervals for bioequivalence trials. *Biometrics* 32(4): 741–744.

Williamson, J., S. Crawford, and H. Lin. 2007. Resampling dependent concordance correlation coefficients. *Journal of Biopharmaceutical Statistics* 17(4):685–696.

Williamson, J., S. Lipsitz, and A. Manatunga. 2000. Modeling kappa for measuring dependent categorical agreement data. *Biostatistics* 1(2):191–202.

Wu, W. 2005. *A Unified Approach for Assessing Agreement, PhD Thesis*. Chicago: University of Illinois.

Yang, M. 2002. *Universal Optimality in Crossover Design and Statistical Methods in Assessing Agreement, PhD Thesis*. Chicago: University of Illinois.

Yang, J. and V. Chinchilli. 2009. Fixed-effects modeling of Cohen's kappa for bivariate multinomial data. *Communications in Statistics–Theory and Methods* 38:3634–3653.

Yang, J. and V. Chinchilli. 2011. Fixed-effects modeling of Cohen's kappa for bivariate multinomial data. *Computational Statistics and Data Analysis* 55:1061–1070.

Zeger, S. and K. Liang. 1986. Longitudinal data analysis for discrete and continuous outcomes. *Biometrics* 42(1):121–130.

Zeger, S., K. Liang, and P. Albert. 1988. Models for longitudinal data: a generalized estimating equation approach. *Biometrics* 44(4):1049–1060.

Zhong, J. 2001. *Optimal and Efficient Nonlinear Design and Solutions with Interpretations to Individual Bioequivalence, PhD Thesis*. Chicago: University of Illinois.

Index

L. Lin et al., *Statistical Tools for Measuring Agreement*,
DOI 10.1007/978-1-4614-0562-7, © Springer Science+Business Media, LLC 2012